Encyclopedia of Mathematics and Society

Engineering, Technology, and Medicine

Encyclopedia of Mathematics and Society

Engineering, Technology, and Medicine

Editors
Sarah J. Greenwald
and
Jill E. Thomley
Applachian State University

SALEM PRESS
A Division of EBSCO Publishing
Ipswich, Massachusetts

Cover Photo: © Amanda Hall/Robert Harding World Imagery/Corbis

Copyright © 2013, by Salem Press, A Division of EBSCO Publishing, Inc.
All rights reserved. No part of this work may be used or reproduced in any manner whatsoever or transmitted in any form or by any means, electronic or mechanical, including photocopy, recording, or any information storage and retrieval system, without written permission from the copyright owner. For permissions requests, contact proprietarypublishing@ebscohost.com.

ISBN 978-1-4298-3753-8

Printed in the United States of America

Contents

List of Contributors   ix

Aircraft Design	1
Airplanes/Flight	4
Animation and CGI	6
Arenas, Sports	8
Auto Racing	9
Bicycles	11
Bridges	13
Calculators in Society	14
Calendars	17
Canals	19
Carpentry	20
Castles	22
Cell Phone Networks	24
City Planning	25
Clocks	28
Coding and Encryption	30
Communication in Society	32
Dams	37
Deep Submergence Vehicles	39
Digital Book Readers	40
Digital Cameras	41
Digital Images	43
Digital Storage	45

Electricity	46
Elevators	48
Engineering Design	50
Fax Machines	52
Fuel Consumption	54
GPS	55
Green Design	57
Helicopters	59
Highways	60
HOV Lane Management	61
Internet	63
Landscape Design	67
Levers	69
Light Bulbs	70
Mapping Coastlines	71
Mattresses	73
Microwave Ovens	74
Movies, Making of	75
MP3 Players	77
Neural Networks	78
Nielsen Ratings	80
Optical Illusions	82
Parallel Processing	84
Personal Computers	85
Pulleys	87
Radio	89
Robots	91
Roller Coasters	94
Search Engines	95
Segway	96
Servers	97
Shipping	98
Skyscrapers	100
SMART Board	102
Smart Cars	103
Social Networks	104

Spaceships	106
Spam Filters	108
Televisions	109
Thermostat	111
Time Signatures	113
Toilets	114
Traffic	116
Trains	117
Traveling Salesman Problem	120
Tunnels	121
Universal Language	123
Vending Machines	124
Video Games	125
Water Distribution	128
Wheel	130
Windmills	131
Wireless Communication	132

Resource Guide — **135**

List of Contributors

Sukantadev Bag
  University College Cork
Zenia C. Bahorski
  Eastern Michigan University
Robert A. Beeler
  Eastern Tennessee State University
Kimberly Edginton Bigelow
  University of Dayton
Vladimir E. Bondarenko
  Georgia State University
Sarah Boslaugh
  Washington University School of Medicine
Chris D. Cantwell
  Imperial College London
Peter J. Carrington
  University of Waterloo
Maria Droujkova
  Natural Math
Gisela Ernst-Slavit
  Washington State University, Vancouver
Daniel J. Galiffa
  Penn State Erie, The Behrend College
Angela Gallegos
  Occidental College
Catherine C. Galley
  Independent Scholar
Mark Ginn
  Appalachian State University
Jeff Goodman
  Appalachian State University
Sarah J. Greenwald
  Appalachian State University
William Griffiths
  Southern Polytechnic State University

Alexander A. Gurshtein
  Mesa State College
Simone Gyorfi
  O. Goga High School, Jibou, Romania
Anthony Harkin
  Rochester Institute of Technology
Jerry Johnson
  Western Washington University
D. Keith Jones
  University of Southampton
David I. Kennedy
  Shippensburg University of Pennsylvania
Matt Kretchmar
  Denison University
Bill Kte'pi
  Independent Scholar
Maria Elizete Kunkel
  University of Ulm, Germany
Alistair Kwan
  Yale University
Carmen M. Latterell
  University of Minnesota, Duluth
Michele LeBlanc
  California Lutheran University
Chad T. Lower
  Pennsylvania College of Technology
Deborah Moore-Russo
  State University of New York, University at Buffalo
Ashwin Mudigonda
  Universal Robotics Inc.
Serkan Ozel
  Bogazici University
Zeynep Ebrar Yetkiner Ozel
  Fatih University

Julian Palmore
University of Illinois at Urbana-Champaign
Robert W. Peck
Louisiana State University School of Music
Michael Qaissaunee
Brookdale Community College
David C. Royster
University of Kentucky
Carl R. Seaquist
Texas Tech University
Shahriar Shahriari
Pomona College
Barbara A. Shipman
University of Texas at Arlington
Lawrence H. Shirley
Towson University
Florence Mihaela Singer
University of Ploiesti, Romania
David Slavit
Washington State University

Mark R. Snavely
Carthage College
Jill E. Thomley
Appalachian State University
Marcella Bush Trevino
Independent Scholar
Juliana Utley
Oklahoma State University
Jiri Wackerman
Institute for Frontier Areas of Psychology and Mental Health
Bethany White
University of Western Ontario
Connie Wilmarth
Northwest Christian University
Todd Wittman
University of California, Los Angeles

Aircraft Design

Category: Architecture and Engineering.
Fields of Study: Geometry; Number and Operations.
Summary: Mathematics plays a pivotal role in designing, manufacturing, and enhancing aircraft components and launch platforms.

Achieving flight has been a dream of mankind since prehistory, one never abandoned. As early as Leonardo da Vinci, mathematics—the cornerstone of engineering and physics—was recognized as the key to realizing the dream. Da Vinci's 1505 "Codex on the Flight of Birds," for instance, is a brief illustration-heavy discussion attempting to discover the mechanics of birdflight in order to replicate those mechanics in manmade flying machines. Da Vinci considered not simply the wingspan and weight of birds but a fledgling notion of aerodynamics. He was the first to note that in a bird in flight, the center of gravity—the mean location of the gravitational forces acting on the bird—was located separately from its center of pressure where the total sum of the pressure field acts on the bird. This fact would be important in later centuries when aircraft were designed that are longitudinally stable. Today, mathematics is used in the study of all aspects of flight, from launch platform design to the physics of sonic booms.

Complex Analysis and the Joukowski Airfoil

Abstract mathematics can find its place in physical applications people experience quite often. For example, complex analysis and mappings play a vital role in aircraft. In layman's terms, complex analysis essentially amounts to reformulating all the concepts of calculus using complex numbers as opposed to real numbers. This formulation leads to new concepts that cannot be achieved with only real numbers. In fact, the very notion of graphing complex functions, rather than real functions, is quite different—mathematicians often call the graphing of complex functions a "mapping." Taking a simplistic geometric figure, like a circle, and then applying a complex function transforms the figure into a more complicated geometric structure. One figure that results from such a transformation looks like an airplane wing. Furthermore, one can consider the curves surrounding the circle as fluid flow, that is, air currents, and we obtain a rudimentary model of airflow around an airplane wing. This transformation is entitled the Joukowski Airfoil, which is named after the Russian mathematician and scientist Nikolai Joukowski (1847–1921), who is considered a pioneer in the field of aerodynamics. Variations of this transformation have been utilized in applications for the construction of airplane wings.

Nature-Inspired Algorithms

An example of how various fields of mathematics, science, and engineering coalesce is epitomized at the Morpheus Laboratory, where applications of methods and systems found in nature are applied to the study and design of various types of aircraft. For example, biologically inspired research is conducted by studying an assortment of details related to the mechanics of birds in flight.

Birds are an example of near perfection in flight, a fact that humans have long observed. Birds have been evolving for millions for years and have adapted to various environmental changes, thus altering their flight mechanics accordingly. By studying the mathematical properties related to their wing morphing, surface pressure sensing, lift, drag, and acceleration, among other aspects, the researchers at Morpheus Laboratory can use the knowledge they have gleaned and apply it to several different types of aircraft. In order to accomplish this feat, mechanical models of actual birds are constructed and analyzed. Morpheus researchers utilize an assortment of mathematics and physics, including fluid mechanics (the study of air flow in this case) and computer simulations, to analyze the data that result from studying the mechanical birds in flight. The analysis, in turn, results in novel perspectives in flight as well as the design of innovative types of planes.

In addition, many of the problems that arise regarding the machinery and components that comprise an aircraft carrier can also be potentially solved via Darwinian-inspired mathematical models. For example, the structural components of aircraft are constantly being optimized, as numerical performance is attempted to be maximized while cost is minimized.

The managing of cabin pressurization has made it possible for aircraft to fly safely under various weather conditions and landscape formations. This ability is due in large part to devices known as "pressure bulkheads," which close the extremities of the pressurized cabins. Because of the wealth of physical phenomena that influence the stability of these bulkheads, such as varying pressures, it has been a challenge to optimize their design. In the early twenty-first century, it was proposed that the bulkheads should have a dome-like shape, as apposed to a flat one, which was suggested by both mathematical and biological evidence. Interestingly, these two structures demonstrate completely dissimilar mechanical behaviors, which lead researchers to consider different approaches to modeling the dome-like bulkheads.

The dome-like structured bulkheads are analogous to biological membranes and can be mathematically modeled in a similar fashion. In addition to the implementation of these membrane-like designs, the minimization of the cost of their construction and the assurance of their durability is mathematically modeled.

Simulating Sonic Booms

Every time an aircraft travels faster than the speed of sound, a very loud noise is produced called a "sonic boom." The boom itself results when an aircraft travels faster than the speed of the corresponding sound waves. The boom is a continuous event, as opposed to an instantaneous sound, which is a result of the compression of the sound waves. Other fast-moving projectiles like bullets and missiles also produce sonic booms.

Mathematically, this concept means that the velocity of an aircraft (v_a) exceeds the wave velocity of sound (v_s). The Mach Number (M), named after the Austrian physicist and philosopher Ernst Mach (1838–1916), is defined as the ratio of the velocity of an aircraft to the velocity of sound. This ratio is expressed mathematically as

$$M = \frac{v_a}{v_s}.$$

When $v_a < v_s$, $M < 1$, the object is moving at what is often referred to as "subsonic speed." If $v_a = v_s$, $M = 1$, and the object is moving at what is frequently called "sonic speed." Whenever $v_a > v_s$, $M > 1$, and the object is moving at what is titled "supersonic speed." Furthermore, whenever $v_a > v_s$, a shock wave is produced.

The shock waves from jet airplanes that travel at supersonic speeds carry a great amount of concentrated energy resulting in great pressure variations. In fact, two booms are often produced when jets fly at supersonic speeds. Usually, these two booms coalesce into an N-shaped sound wave that propagates in the atmosphere toward the ground. Although shock waves are exceedingly interesting, they can be unpleasant to the human ear and can also cause damage to buildings including the shattering of windows.

However, there is increasing economical interest in designing aircraft carriers that can travel at supersonic speeds with a low sonic boom. To demonstrate, the flight time for a trip from New York to Los Angeles can essentially be cut from 10% to 50% if the plane flies at a supersonic cruise speed instead of subsonic speed. Therefore, physicists are currently developing adaptive methods that model sonic booms in order to ultimately develop aircraft that can travel at supersonic speeds without causing structural damage—aircraft that create a low sonic boom. Aspects such as near-field airflow as well as pressure distribution have been analyzed in these models by utilizing techniques of mathematical analysis.

Aircraft Carriers

Airplanes were a major evolution in modern warfare. World War II aircraft carriers that moved airplanes closer to targets that would otherwise be well beyond their fuel ranges proved to be pivotal to many battles, especially in the Pacific. They continue to be a key component of many countries' navies for rapid deployment of aircraft for surveillance, rescue, and other military uses. Launching from and landing airplanes on aircraft

Scientists at the U.S. National Aeronautics and Space Administration (NASA) envisioned this design as a twenty-first-century aerospace vehicle. The "Morphing Airplane" is part of NASA's vision for aircraft of the future. (National Aeronautics and Space Administration)

carriers is considered one of the most challenging pilot tasks because of the restricted length of the deck and the constant motion of the deck in three dimensions. A catapult launch system gives planes the added thrust they need to achieve liftoff and requires calculations that take into account mass, angles, force, and speed. Similar issues apply to the tailhook capture system that stops planes when they land.

There are also significant scheduling issues for multiple aircraft on a carrier, fuel use, weapons logistics, and radar systems used to monitor both friendly and enemy planes. Aircraft carriers are like large, self-contained floating cities. Mathematicians work in the nuclear or other power plants that provide electricity for the massive aircraft carriers of the twenty-first century and in many other logistics areas beyond direct flight launch and control. They also help design and improve aircraft carriers. For example, mathematician Nira Chamberlain modeled the lifetime running costs of aircraft carriers versus operating budgets to develop what are known as "cost capability trade-off models," which were used to help make decisions about operations. He also worked on plans for efficiently equipping ships to optimize speedy access to spare components. Some of the mathematical methods he used include network theory, Monte Carlo simulation, and various mathematical optimization techniques.

Further Reading

Alauzet, Frederic, and Adrien Loseille. "Higher-Order Sonic Boom Modeling Based on Adaptive Methods." *Journal of Computational Physics* 229 (2010).

Balogh, Andres. "Computational Analysis of a Boundary Controlled Aircraft Wing Model." *Sixth International*

Conference on Mathematical Problems in Engineering and Aerospace Sciences. Cambridge, England: Cambridge International Science Publishing, 2007.

Freiberger, Marianne. "Career Interview With Nira Chamberlain: Mathematical Modelling Consultant." http://plus.maths.org/content/career-interview-mathematical-modelling-consultant.

Morpheus Laboratory. http://www.morpheus.umd.edu.

Niu, Michael Chun-Yung, and Mike Niu. *Airframe Structural Design: Practical Design Information and Data on Aircraft Structures.* Granada Hills, CA: Adaso Adastra Engineering Center, 1999.

Viana, Felippe, et al. "Optimization of Aircraft Structural Components by Using Nature-Inspired Algorithms and Multi-Fidelity Approximations." *Journal of Global Optimization* 45 (2009).

Yong, Fan, et al. "Aeroservoelastic Model-Based Active Control for Large Civil Aircraft." *Science China: Technological Sciences* 53 (2010).

Daniel J. Galiffa

Airplanes/Flight

Category: Travel and Transportation.
Fields of Study: Algebra in Society; Geometry in Society; Number and Operations.
Summary: Aerodynamics is necessary to understanding the flight of objects through three-dimensional space and the forces acting upon them.

Human flight involves moving in a three-dimensional environment within the atmosphere in a stable, controlled way. Aerodynamics is the study of forces and the resulting motion of objects through air. It comes from Greek *aerios*, meaning "air," and *dynamis*, meaning "force." Mathematics is fundamental to understanding flight and in the design of different flying devices and machines, including kites, balloons, helicopters, and airplanes. From Orville and Wilbur Wright's initial experiments with gliders at the beginning of the twentieth century, to the breaking of the sound barrier in the middle of the century, to the development of suborbital craft at the start of the twenty-first century, airplanes have been constructed in many different forms.

However, the ability to fly for all fixed-wing aircraft ultimately depends on a differential movement of air above and below the wings to generate positive lift. Control depends on three parameters, known as "pitch," "yaw," and "roll," that are angles of rotation in three dimensions or axes about the plane's center of mass. Mathematicians and others continue to study flight in order to more fully understand the mathematical and scientific principles that keep heavier-than-air craft in the air and to produce designs that are faster, safer, and more efficient. They also explore related issues in air travel, such as optimal strategies for loading passengers onto planes and the scheduling of aircraft flight crews.

Mathematical History

Stories from many cultures around the world suggest that humans have been interested in flight for thousands of years. There is evidence that the Chinese used kites well before the first century C.E. Leonardo da Vinci recorded his studies of flight in the fifteenth century with more than 100 drawings, including his theoretical ornithopter. Air is a fluid, and so much of the mathematics of flight science derives from fluid force studies, such as those performed by mathematician Daniel Bernoulli in the seventeenth century. Bernoulli's principle is one foundation of flight mechanics.

Mathematical models for flight rely on the Navier-Stokes equations, named for mathematicians Claude-Louis Navier and George Stokes, which are fundamental partial differential equations describing fluid flow. They have many extensions. The Darcy–Weisbach equation, derived by dimensional analysis and named for engineer Henry Darcy and mathematician Julius Weisbach, is important to understanding the dissipation of energy because of friction, such as drag. Working in the early twentieth century, mathematician Otto Blumenthal studied the theory of complex functions, which he also applied to problems such as stress in airplane wings. Mathematician Selig Brodetsky studied equations of airplane motion, including three-dimensional phugoids, which are extensions of common, undesirable oscillatory motions where a plane pitches up and climbs, then pitches down and descends, with changes in airspeed. Peter Lax studied a class of nonlinear equations that can develop singularities, which have applications in aerodynamics that are related to phenomena like the shock waves that result from breaking the sound barrier.

Principles of Flight

Balloons are an example of lighter-than-air craft that use buoyancy to ascend and descend within the atmosphere, and hot air balloons are known to have been explored and used in the eighteenth century. There is also evidence that miniature hot air balloons were used in China for several centuries.

Heavier-than-air craft use the principle of lift to overcome gravity. There have been various mathematical and physical theories posed regarding how lift in airplane wings is accomplished. Aerodynamicists have analyzed how the motion of the air over an airplane wing creates circulation and differential pressure above and below the wing, which creates lift. Lifting forces on the airfoil are perpendicular to the motion of the lifting surface through the air and, in level flight, they counteract gravity. An observable example is the "sing" or hum that occurs in telephone wires in a steady wind, which is a repeating pattern of swirling vortices. This effect is because of the oscillations induced by a phenomenon called "vortex shedding," which causes the wires to oscillate perpendicular to the wind flow.

Studies and models suggest that an airfoil produces circulation in a similar manner. Airfoils can be optimally designed to take advantage of this effect by allowing a smooth flow to develop over the surface of the airfoil, called "laminar flow." The Reynolds number, named for mathematician Osborne Reynolds, quantifies laminar flow. Without laminar flow over an airfoil, turbulence is produced and vortex shedding occurs. Others suggest that aircraft lift is a Newtonian reaction force, named for Isaac Newton, coupled with the Coandă effect, named for engineer Henri Coandă, which is the tendency of a fluid to be attracted to a surface, like an airplane wing. The wing pushes the air down, so the air pushes the wing up.

Lift and Thrust

In general, a pilot taking off from the ground initially accelerates directly into oncoming wind whenever possible, since there is agreement based on observation and mathematics that relative forward motion of the plane's wings with respect to the air is required for flight. Usually, the plane itself is in motion, though a strong wind over a stationary wing can also generate some lift. To maintain a steady, level flight path after takeoff, without any added acceleration, two mathematical relations must be maintained: thrust = drag and lift = weight. Early aircraft engines were powered by gasoline, similar to automobile combustion engines. A fundamental problem of weight, which inhibited lift, was solved by using aluminum as a construction material. Although oxygen is needed to burn gasoline, it is not carried by the aircraft but extracted from the atmosphere so that it does not add to the mass of the aircraft. Jet engines compress and discharge a fast-moving jet of air to generate thrust, using the same principles of fluid dynamics that govern other aspects of aircraft flight, according to Newton's third law of motion. In contrast, a rocket must carry propellants, both fuel and oxidizer, and can thus fly outside of the atmosphere. The added force helps compensate for the extra weight.

Flight Speed

The types of speeds of flight are typically classified as slow subsonic flight, fast subsonic flight, trans-sonic

The Father of Aviation

Engineer George Cayley (1773–1857), working in the eighteenth and nineteenth century, is often called the "father of modern aviation" for his research, which helped identify the aerodynamic forces of flight: weight, lift, drag, and thrust. Though Cayley experimented with manned gliders, modern heavier-than-air flight is generally traced to the 1903 launch of the Wright Flyer, a twin propeller biplane with a single motor to provide thrust and mechanisms so that the pilot could control for pitch, roll, and yaw.

Their design helped overcome previous obstacles to sustained stable and controlled flight by adding ailerons to the wings, elevators to the tail surfaces, and rudders to the fuselage to manage airflow. By common convention, roll is motion about the longitudinal axis of the plane. Yaw is movement about the vertical body axis. Pitch is movement about an axis that is perpendicular to the longitudinal plane of symmetry. Pilots require a firm grasp of this three-dimensional geometry to navigate aircraft and follow directional headings.

flight, and supersonic flight. The Bell X-1 rocket-propelled airplane is credited as the first piloted aircraft in the world to break the sound barrier, under control of test pilot Charles Yeager. Other planes have been thought to have broken the sound barrier during steep dives, which many do not consider flight. The joint United Kingdom and France plane known as the *Concorde*, which flew from the 1970s until its retirement in 2003, was the only commercial supersonic aircraft. Commercial jets of the early twenty-first century typically achieve speeds in the range of 80% to 85% of the speed of sound, the slower end of trans-sonic flight.

The design speeds tend to avoid compressibility effects in air, which occur above roughly 80% of Mach 1. The Mach number is a ratio of the speed of the aircraft to the speed of sound at the aircraft's altitude. Supersonic flight requires much more energy to sustain, and generally only military aircraft conduct sustained supersonic flight within the atmosphere. The Prandtl-Glauert equation, named for scientists Ludwig Prandtl and Hermann Glauert, is used to help correct computations of fluid flow at high speeds a function of compressibility, while the Prandtl-Glauert singularity is observed as a visible cloud of vapor that results from air pressure changes around a trans-sonic airplane. The pressures can be modeled as an N-wave, named because a plot of pressure versus time resembles the letter N.

A mode of atmospheric flight explored with experimental aircraft at the beginning of the twenty-first century is hypersonic flight, which starts at speeds approximately 5–10 times the speed of sound. Special engines must be developed to make this speed possible. Previously, the Lockheed Aircraft SR-71 held the speed record at greater than Mach 3. It was powered by a special fuel and was air breathing. In 1974, the SR-71 set a speed record flying across the Atlantic from Beale Air Force Base in Louisiana to London in less than two hours. This flight occurred many decades after aviator Beryl Markham's speculations about flying the Atlantic in an hour. Hypersonic aircraft flying at speeds greater than Mach 5 likely will be powered by different forms of air breathing propulsion systems, such as turbine-free engines known as "scramjets," which at very high speeds use ram air compression to ignite a fuel in the engine. In principle, such designs have the capability of going at very high speeds at high altitude and form a transition to spaceflight.

Further Reading

Anderson, David, and Scott Eberhardt. *Understanding Flight*. 2nd ed. New York: McGraw-Hill, 2009.

Tennekes, Henk. *The Simple Science of Flight: From Insects to Jumbo Jets*. Revised and expanded ed. Cambridge, MA: MIT Press, 2009.

JULIAN PALMORE

Animation and CGI

Category: Arts, Music, and Entertainment.
Fields of Study: Geometry; Measurement; Representations.
Summary: Animators have become adept at creating realistic products with the help of mathematics.

Animation is the process of creating the illusion of fluid movement from a series of static images. When these images are viewed in sufficiently fast succession, the human eye sees them as continuous motion rather than a sequence of discontinuous still images. From the earliest mechanical devices, through hand-drawn, stop-motion, and computer-assisted film techniques of the twentieth century, up to the latest computer-generated imagery (CGI), the quest has been to create interesting representations of movement and action.

Early Animation Devices

Historically, there have been several mechanical devices that were developed to simulate movement using still pictures. The Phenakistoscope, invented in 1832 by the Belgian physicist Joseph Plateau, consisted of a spindle with two mounted discs, one with slots around its edge and the other with pictures of successive action. With the discs spinning in unison and the picture side facing a mirror, the view through the slots appeared to show a moving drawing. In 1834, a British mathematician named William Horner produced the Zoetrope, a cylinder cut with vertical slits. Pictures of successive action were positioned on the inside opposite the slits. With the cylinder rotating, the image seen through the slits appeared to be in motion. As the Zoetrope used more pictures, and could be rotated more quickly, this gave a better illusion of movement. Even in the early twenty-first century, the Zoetrope is used to illustrate the basic idea of animation.

Animation Principles

By the start of the twentieth century, these mechanical devices were superseded by animated films. The principal technique was to hand-draw each frame. In the 1930s, animators at Walt Disney Studios developed what became known as the "12 principles of animation," many of which remain pertinent in an era of CGI. To illustrate, consider someone throwing a ball so that it bounces along the ground. A thrown ball is known to follow a parabolic path, a form of arc. The "arc" principle of animation is that almost all actions follow some form of arc. Arcs, as the Disney animators were well aware, give animation a more natural appearance. Another principle, "slow in and slow out," relates to the ball taking time to accelerate and decelerate. The animation looks most realistic if there are more frames near the beginning and end of a movement, and fewer in the middle. The flight of the ball and its bounce involves the principle of "squash and stretch." As the ball falls, animating a slight stretch gives the impression of the ball having speed. Dilation is the mathematical transformation for stretching and shrinking. Animating a squash to the ball as it bounces gives the impression of weight. For the ball to seem real, the animator uses the principle of "solid drawing" by taking into account the form of the ball in three-dimensional space as well as using the geometry of light and shadow.

CGI and Mathematics

CGI is even more mathematically based than hand animation because the images must be mathematically represented in order to be manipulated in the computer environment. Oscar-winning computer scientist Tony DeRose, who has worked for Pixar Animation Studios, said, ". . . different kinds of mathematics are used for different aspects of a film, from the simulation of how light bounces around in an environment (integral calculus) to obtaining smooth surfaces efficiently (subdivision surfaces) and making characters move in a realistic fashion (harmonic coordinates)." Trigonometry and vector algebra are widely used in CGI algorithms for creating and manipulating images. Matrices are a standard algebraic way of representing various transformations. Dilation makes objects larger or smaller in addition to stretching; translations move objects; and rotations turn objects.

One classic CGI method for creating three-dimensional animated objects involves using polygonal meshes, which are collections, or grids, of polygons. This method makes use of the geometry of smooth surfaces. Like animated motion, this method relies on the human eye's tendency to smooth discontinuous regions. Locally, smooth surfaces look flat, so they can be approximated with small, flat polygons such as triangles or quadrilaterals. Basic three-dimensional shapes such as cubes, cylinders, spheres, and cones may be joined to form composite three-dimensional objects. Interpolation is also used. More complex and smoother-looking three-dimensional objects can be modeled using sophisticated mathematics like spline patches and nonuniform rational basis splines, where a spline is a mathematical function defined piecewise by polynomials. Such techniques have become standard practice in CGI. The mathematical representation of three-dimensional shapes, including layout and materials, is used to compute a two-dimensional image from a given viewpoint, a process called "rendering." This process entails addressing issues such as visibility from selected viewer angles (including which parts of objects in the scene are hidden) and appearances, and how objects look different as the lighting varies. Finally, the motion of each object in the scene has to be specified.

Lucasfilm LTD and animator Kecskemeti B. Zoltan of Ste-One provided mathematician Timothy Chartier with digital models of Yoda from the *Star Wars* movies to explore in linear algebra classrooms. One of the models had 53,756 vertices, 4040 triangles, and 49,730 quadrilaterals, illustrating that realistic images and their transformations have many more data points and matrix multiplications than is typical as classroom examples. Chartier noted, "More recently, computer animation produced the character's movement, which required mathematical concepts from such areas as linear algebra, calculus, differential equations, and numerical analysis. Drawing on these popular culture ties in appropriate coursework can pique students' curiosity and compel further learning."

Despite the many available mathematical techniques and advances in the computational and visualization power of computer systems, convincing simulation of some physical features, like hair, continues to be challenging. Pixar noted that it took up to 12 hours to render a single frame of the character Sulley in the 2001 movie *Monsters, Inc.* because of his nearly 3 million individually animated hair strands. Each hair was mathematically modeled as a series of springs connected via hinges.

CGI has come a long way since the 1976 movie *Futureworld*, which many acknowledge as the first use of three-dimensional computer imagery. Even though the first CGI film to win an Oscar was Pixar's short movie *Tin Toy* in 1988, the 1995 movie *Toy Story* was the first full-length, fully CGI feature film. Many challenges and problems remain to be solved in the quest for photo-realism in CGI. Examples include more accurate modeling and representation of physical actions, such as swallowing, as well as textural and other properties of materials like skin, including wrinkles. Animators also seek to better differentiate faces for people of varying ages, such as children or the elderly.

Further Reading

Chartier, Timothy. "Using the Force: Star Wars in the Classroom." *PRIMUS* 17, no. 1 (2007).

Mathematical Association of America. "Pixar's Tony DeRose Illuminates the Mathematics of Animation." http://www.maa.org/news/101509derose.html.

McAdams, A., S. Osher, and J. Teran. "Crashing Waves, Awesome Explosions, Turbulent Smoke, and Beyond: Applied Mathematics and Scientific Computing in the Visual Effects Industry." *Notices of the American Mathematical Society* 57, no. 5 (2010). http://www.ams.org/jackson/fea-mcadams.pdf.

D. Keith Jones
Deborah Moore-Russo

Arenas, Sports

Category: Games, Sport, and Recreation.
Fields of Study: Algebra; Geometry.
Summary: Modern arena designers consult mathematicians to determine the effects of design on play and crowd behavior.

A sports arena is essentially an enclosed area consisting of a large open space where a sport is played, surrounded by seating for spectators. It may also include various facilities for athletes, spectators, and the press. Many sports use specific terms for arenas, like "park" for baseball and "stadium" for football. Some sports arenas are open-air while others are roofed. The word "stadium" comes from *stadion*, an ancient Greek unit of length. Mathematics plays a significant role in the design and maintenance of modern sports arenas, including not only the geometrically shaped playing surfaces but also the optimization of seating, sightlines, acoustics, lighting, spectator traffic flow, and placement of restrooms and concessions. Features such as retractable roofs and convertible forms to accommodate multiple sports require careful design as well. Mathematicians also analyze and model features of sports arenas to determine their potential effect on the game play.

The rules of each sport dictate dimensions for the field of play. Some such as hockey, football, basketball, and soccer specify exact dimensions for the playing surface and delineate areas for specific activities, like the rectangular key in basketball or the half-circle goal crease for amateur hockey. Baseball, on the other hand, standardizes the dimensions of some features such as the distance between bases and the distance between the pitcher and home plate, but the outfield varies depending on the positions of the outfield walls. Further, aspects of game play can be affected by design choices. Fenway Park's outfield wall known as the "Green Monster" is notorious for stopping home runs, yielding more doubles and triples. When the new Yankee Stadium produced a higher rate of home runs, there was speculation about a "wind tunnel" effect. Statistical analyses suggested that curvature and height of the right field wall were more important than wind speeds or patterns. Statistician George "Bill" James developed the concept of park factors, which attempts to measure how park characteristics influence game outcomes.

Robert F. Kennedy Memorial Stadium in Washington, D.C., which opened in 1961, was the first multiuse stadium. It was widely decried for being a "concrete donut." Some critics suggested its wavy shape and curvature optimized it for baseball seating, though the widely replicated design has deficiencies for baseball and football. Some critical seats were too low for football and too high for baseball, resulting in poor sightline angles.

The baseball configuration was also more symmetrical than most baseball-only fields. Modern designers use mathematical techniques and tools (such as Mathcad software), simulations, and three-dimensional modeling for their designs, resulting in unique facilities like The Float in Singapore, which is literally floating on Marina Bay. Similar methods are involved in the design of arena roofs or domes, some of which are retractable. Calculating the amount of material needed to construct a curved dome, as well as calculating the weights, forces, and stresses, typically involves the use of calculus. These calculations, in turn, partially determine the type of support required.

Geometry and graph theory also contribute to dome design. R. Buckminster Fuller suggested that domes are strongest when the edges lie along great circles. Triangles are often used to give great strength with minimal weight, while other support structures resemble the latitude and longitude configuration on a globe. Fibonacci sequences and plane tilings also are used in the design of some domes. Veltins Arena in Germany uses features like hinged columns with ball-bearing edges that move in three dimensions. Both Veltins Arena and University of Phoenix Stadium in the United States feature sliding roofs and retractable natural-grass playing surfaces weighing millions of pounds. These were mathematically modeled extensively before construction. Transformative structures of this type have become known as "kinetic architecture."

Mathematicians continue to investigate questions related to sports arenas, some of which have wider applications. Researchers have considered the impact of sports arenas on land values using hedonic regression models. Mathematical analyses of crowd sequence videos (frequently taken from sports venues) benefit research in areas including surveillance, designs of densely populated public spaces, and crowd safety. In some cases, people are conceptualized as a "thinking fluid" to which fluid dynamic and stochastic models may be applied. Unusual events like emergency evacuations are fairly rare, and there are legal barriers to obtaining extensive live footage. As such, computer scientists and mathematicians have developed detailed simulations for both "normal" behavior and unusual crowd events. Some have suggested that topology optimization would be beneficial for investigating arena evacuation plans.

Further Reading

Puhalla, Jim, Jeffrey Krans, and Michael Goatley. *Sports Fields: Design, Construction, and Maintenance.* Hoboken, NJ: Wiley, 2010.

Winston, Wayne. *Mathletics: How Gamblers, Managers, and Sports Enthusiasts Use Mathematics in Baseball, Basketball, and Football.* Princeton, NJ: Princeton University Press, 2009.

BILL KTE'PI

Auto Racing

Category: Games, Sport, and Recreation.
Fields of Study: Data Analysis and Probability; Geometry; Measurement.
Summary: Mathematics is essential in the design of race cars and racetracks, and the formulation of race strategy.

Auto racing has taken place for as long as cars have existed. While the early days of racing were related to fairly simple vehicles, it is now a very technical sport that has multiple branches with fans worldwide. Auto racing includes not only cars that are similar to those driven by the average citizen but also cars that are very sophisticated. The different branches of auto racing differ in the specifics of the car but all share a strong relationship to mathematical principles. The design of the car, its tires, the track, and the drivetrain require very careful measurement. The optimal path for a given track and weather condition requires a deep understanding of angles and geometry. Analysis of data to create probability information enables drivers and their teams to make wise decisions for a given set of conditions during a race.

Overview

Auto racing began as soon as the automobile was invented in the late 1800s. Auto racing is a broad term that includes single-seat cars or open wheel cars, which the Indianapolis 500 has made famous. Formula 1 racing is another type of open wheel racing but involves racing around courses that are not oval shaped. The National Association for Stock Car Auto Racing

Drivers preparing during a practice run for the 2004 Daytona 500 race. An understanding of geometry is critical in determining how to set up cars to handle banking and high speeds. (U.S. Air Force, Larry McTighe)

(NASCAR) utilizes cars that are modified from cars that can be bought by the general public. Many successful professional race car drivers began their racing careers with Kart racing, which involves vehicles that look like sophisticated go-karts.

Race Track Design

The racing surface and the track design are significant factors that affect both car design and driving strategy. Race surfaces can include asphalt, concrete, dirt, sand, and (sometimes) ice. Some tracks consist of a very short distance (1/4 mile) and are straight. These tracks are typically used for drag racing, which involves cars trying to go as fast as possible over a short distance. Many track designs have drivers travel in an oval, or near-oval, shape with some banking to help make high-speed turns easier. An understanding of geometry is imperative when determining how best to set up the car to handle the banking and the speeds.

Track designs also include road courses in which racers turn both left and right and require a completely different car design to handle banking in both directions. The radius of a turn influences how fast the car can go without losing grip and crashing into the outer wall. The speed affects the size of the down force on the car (caused by spoilers), and, as such, different tracks require car designs.

Car Design

Race car designs evolve in response to technology changes and safety concerns, often as a result of mathematical or statistical analysis. Each branch of auto racing has very strict rules on car design, which are tested before—and sometimes after—each race. The testing includes very careful measurements of various components of the car from the size of various components of the engine, to the car's width, height, and weight. The tests focus on items that affect the car's power (the engine), response to the environment (temperature, air resistance, and gravity), and its influence on forces that are made on the car (width, height, and weight). Because they are such an important part of car performance, tires are supplied to the teams. A large amount of testing by tire companies goes into determining which type of tires will be provided for a particular track. The air and track temperatures often change drastically during a race and can affect how the tires interact with the track surface—providing more or less grip. Likewise, the gas that is put into the car is also provided to drivers. These standardizations provide a more even playing field for the teams so that the driver who wins is, presumably, the one with the greatest skill. Teams can alter the cars slightly during races to modify how the car receives forces from the track and from the air. These modifications include taking out or adding small wedges that alter the angle that the car sits on the track. The impact these small changes make on force is understood using trigonometry.

Race Strategy

Once teams have prepared their car and driver for the race, the issue of strategy plays an important role. Teams use probabilities to determine if and when to stop in the pits to change tires or to add gas. Gas mileage is estimated by using regression involving the number of laps, the speed of the car during the laps run, and the temperature. This estimation is not absolutely exact, and it is not uncommon for drivers to run out of gas near the end of some races because of an error in the team's regression model. Some teams alter the usual pit stop, which involves replacing all of the tires and adding gas, by replacing just some of the tires or just adding gas.

Technology and Safety

Technology is playing a bigger role in auto racing in both car development and car testing. Car teams now use technology to measure a large number of factors that influence their car's performance. For some branches of auto racing, these measurements are made during races. For other branches, the rules prohibit this during races but allow the measurements to occur during practice and research design. Because testing can be so expensive, some tests are done with a few drivers and then shared with all the teams. The use of computer simulation based on mathematical modeling is becoming more prevalent in all branches. It is not unusual for teams to use wind tunnels to test car design, and fluid dynamic modeling has been used to improve the aerodynamic properties of race cars. Off-season drivers use sophisticated driving simulators to hone their skills.

Technology has also been used to make racing safer. Race uniforms, helmets, and car interiors have become much less dangerous because of technological improvements. Additionally, track walls now include what is called a Steel and Foam Energy Reduction (SAFER) barrier, which dissipates the collision energy from a crash so that the impact force felt by the car and driver is smaller and less dangerous.

Further Reading

Beckman, Brian. "The Physics of Racing." http://phors.locost7.info/contents.htm.

Bentley, Ross. *Speed Secrets: Professional Race Driving Techniques*. Osceola, WI: MBI Publishing, 1998.

Genta, Giancarlo. *Motor Vehicle Dynamics: Modeling and Simulation*. Singapore: World Scientific Publishing, 1997.

Gifford, Clive. *Racing: The Ultimate Motorsports Encyclopedia*. Boston: Kingfisher Publications, 2006.

Leslie-Pelecky, Diandra L. *The Physics of NASCAR: The Science Behind the Speed*. New York: Plume, 2009.

Michele LeBlanc

Bicycles

Category: Travel and Transportation.
Fields of Study: Algebra, Geometry.
Summary: Bicycle geometry impacts performance, aerodynamics, efficiency, and stability.

The first bicycles of the early nineteenth century were simple designs of wooden frames and metal hoops for wheels. Though these early bicycles were propelled by feet pushing along the ground, soon pedals were added to the front axle allowing the rider to drive the front wheel for locomotion. It was not until the late 1880s when the first chain-driven bicycle was introduced, thereby separating the axles from the primary point of locomotion and overcoming problems with handling, steering, and weight distribution. This explosive decade of development also saw the first pneumatic tires, gear-ing, and coaster brakes, the latter allowing the rider to brake by pedaling backwards. Another series of innovations a century later was spurred by an explosion in frame design and fabrication techniques including the use of better materials such as aluminum, titanium, and, eventually, carbon fiber.

Bicycles serve as the primary means of transportation in several cultures, especially in southeast Asia. European communities are also known for embracing the bicycle as a legitimate form of transportation.

Mechanics

Bicycles have two in-line wheels and are driven by pedaling. The wheels each spin on axles rotating on bearing surfaces and most commonly support the rims via tension spokes. Pneumatic tires are secured to the outer surface of the rims to provide the primary contact with the ground. The centrally located bottom bracket is the rotating connection point of the pedals. Power is transferred to the rear wheel via a chain. Brakes are usually found on both wheels; most bicycles' brakes squeeze braking pads on the rim surface to create friction and slow the wheel and, as a result, the bicycle. Many newer mountain bicycles use disc brakes for increased stopping power. The rider sits on a saddle atop the bicycle and leans forward on handle bars, which provide support and the ability to steer. Many bicycles, especially mountain bicycles, have shock absorbers built into the front fork to provide cushioning over rough terrain. Some bicycles also feature rear suspension, which allows the rear triangle of the frame to rotate and further absorb the impacts of uneven terrain.

Gears (chain rings on the bottom bracket, a cassette on the rear axle) allow the rider to alter the ratio of pedal rotation to wheel rotation in order to go faster or slower. The gear ratio is determined by the diameter of the chain ring divided by the diameter of the rear cog. Since the number of teeth is proportional to diameter, tooth count is more typically used. For example, a 39-tooth chain ring used with a 15-tooth cog produces a gear ratio of

$$\frac{39}{15} = 2.6$$

that is, one revolution of the pedals produces 2.6 revolutions of the rear wheel. A standard 700C wheel (70 centimeters in diameter) will travel $0.7\pi = 2.2$ meters (7.2 feet) along the road with each revolution. Thus, a single rotation of the pedals produces

$$\frac{39}{15}(0.7)\pi = 5.7 \text{ meters (18.7 feet) of travel.}$$

Speed and distance traveled can then be calculated based upon the rider's revolutions per minute.

Types of Bicycles

Reflecting their wide versatility, bicycles come in a multitude of different styles. One of the most common is the road bicycle, which is distinguished by thin tires; a drop-style handlebar; and a stiff, light frame. Road bicycles are designed for fast travel over smoother road surfaces. The other most common bicycle is the mountain bicycle, which features wide, knobby tires designed for increased traction in the dirt; flat handlebars for a more upright position; and a wide range of gears, including very low gears for steep climbing. Most mountain bicycles have a front suspension fork and many feature a rear suspension as well.

Cyclocross bikes are closely related to road bikes but have slightly wider tires and lower gears for racing on cyclocross race courses or for exploring gravel roads. Comfort bicycles, commuters, and hybrids are usually compromises between the stiffness of a road bicycle and the comfort of a mountain bicycle; these bicycles' lower prices are often aimed at entry-level riders who are seeking practicality over high performance. Bicycle motocross (BMX) bicycles are single speed (no gears) with smaller, wider tires designed for racing on BMX courses. There are additional niche bicycles for special purposes such as time trialing, track racing, snow riding, and touring. Though most people cannot imagine a bicycle having anything but circular wheels, since that shape travels smoothly on flat roads, mathematicians have modeled as well as built wheels with other shapes, such as squares, three-leaf clovers, star-like shapes, and triangles. They found that a square-wheeled bike will travel smoothly on a road made of inverted catenaries, and each of the other types has at least one solution as well. A differential equation can be used to generally solve the problem of noncircular wheels.

Racing and Performance

Bicycle racing is a popular sport with a surprisingly active history. Near the end of the nineteenth century, bicycle racing was one of the most popular sports, drawing huge crowds of spectators across Europe and the United States. Today, bicycle racing is popular worldwide but has a stronger European following. Why certain cyclists are more successful than others can be analyzed in part using mathematics. Average riding speed, efficiency, and power are all calculated metrics useful for assessing performance. Seven-time Tour de France winner Lance Armstrong has been studied and modeled extensively throughout his career. American cyclist Greg LeMond overcame a 58-second deficit and won the 1989 Tour de France by 8 seconds over French favorite Laurent Fignon, which is generally attributed by most to the innovative aerodynamic handlebars he used in the last stage. Companies now routinely use mathematical modeling for cycling equipment, as well as to test aerodynamics and other essential properties, and teams use optimization strategies to construct bicycles within the sport's guidelines, since seconds can make the difference between victory and second place.

For the average rider as well as for professionals, the geometry of a bicycle plays a large role in its overall performance and stability. For example, the distance between the axles and the angle the front fork makes with respect to the ground are both important, according to bicycle makers. Some mathematicians have explored stability issues. In study released in 2007, researchers investigated and dynamically modeled 25 parameters believed to be important, with the goal of being able to construct bicycles targeted toward riders' specific needs.

Further Reading

Burke, Ed. *High-Tech Cycling*. 2nd ed. Champaign, IL: Human Kinetics, 2003.

Hall, Leon, and Stan Wagon. "Roads and Wheels." *Mathematics Magazine* 65, no. 5 (1992).

Herlihy, David. *Bicycle: The History*. New Haven, CT: Yale University Press. 2006.

McGann, Bill. *The Story of the Tour de France*. Indianapolis, IN: Dog Ear Publishing, 2006.

Peveler, Will. *The Complete Book of Road Cycling and Racing*. New York: McGraw-Hill, 2008.

Matt Kretchmar

Bridges

Category: Architecture and Engineering.
Fields of Study: Algebra; Geometry.
Summary: Bridges are subject to various complex forces, the distribution of which are determined by their structures.

Bridges are structures built to span a gap or a physical obstacle such as a road or body of water. The many forces acting on bridges make different designs variously suited to different conditions, uses, and building materials. The earliest manmade bridges emulated naturally occurring bridges, like fallen trees that spanned rivers, and were improved upon by lashing logs into place, cutting planks to form a more even travel surface, and eventually building bridges out of stone. The mathematics of bridges was not well understood and most improvements were achieved through trial and error, one of the most significant being the advent of the arch bridge, introduced in Greece in 1300 B.C.E. and used extensively by the Romans. Arch bridges use arch-shaped abutments, sometimes in a series, to distribute much of the bridge's load into horizontal thrust the abutments can restrain—not only a major improvement over earlier designs, but a design well-suited to the simple building materials of the time as stone is strong in compression but weakly resists tension. As applied mathematics became more sophisticated, bridges were often objects of study.

Most bridges are built for functional purposes, but some of them are works of art, like the Golden Gate Bridge or the London Bridge. Mathematicians have long worked on various aspects related to the design and construction of bridges. For instance, Charles Hutton worked on equilibrium principles and Claude-Louis Navier developed a theory for suspension bridges. Applied mathematician P. Joseph McKenna analyzed bridge oscillations and differential equation models of the collapse of the Tacoma Narrows Bridge. The configuration of bridges in Konigsberg served as the subject of mathematical study for Leonhard Euler and is sometimes noted as the beginning of graph theory.

Types of Bridges

There are various types of bridges. Beam bridges consist of a horizontal beam with two supports called "piers" at the ends. Arch bridges are one of the oldest types of bridges and distribute the load of the bridge outward along the curve of the arch to the supports at the ends. Suspension bridges are light and strong and can span longer distances than any other type of bridge, but they are expensive to build. Large bundles of cables suspend the roadway from one end of the bridge to the other. Early Asian suspension bridges were suspended with bamboo cables. Cable-stayed bridges look like suspension bridges, but their cables are secured to towers that bear the load of the bridge. They cost less and their construction is faster than suspension bridges, since they need fewer cables and builders can use pre-cast concrete sections. Movable bridges can be occasionally levered for making way for ships or other kinds of traffic. Double-decked bridges have two levels and are used for multiple forms of traffic—subway, pedestrian, automobile, or bicycle.

The Seven Bridges of Konigsberg

Mathematician Leonhard Euler posed the problem of the seven bridges of Konigsberg in a 1736 paper. The town of Konigsberg contained an island with two branches of a river flowing around it. There were seven bridges spanning the river, and the question was whether a person could start at some point and follow a path that would cross each bridge exactly once and return to the starting point. Euler proved that there was no such path.

Famous Bridges

Millau Bridge, France, is 1125 feet high—higher than the Eiffel Tower. Hangzhou Bay Bridge, China, is 22 miles long. The Rolling Bridge, England, is 39 feet long and rolls itself up until the two ends meet, using a hydraulic press. Tower Bridge, England, is a landmark of London and opens in the center, allowing ships to sail through. Ponte Vecchio, Italy, is considered by some to be the oldest stone arch bridge in Europe. Lake Pontchartrain Causeway, Louisiana, is 24 miles long. Vasco da Gama Bridge, Portugal, is 10.5 miles long. Confederation Bridge, Canada, is 8 miles long. Golden Gate Bridge, California, is one of the most famous symbols of San Francisco. Evergreen Point Floating Bridge, Washington, is a 1.5-mile-long floating bridge.

Further Reading

Blockley, David. *Bridges: The Science and Art of the World's Most Inspiring Structures.* New York: Oxford University Press, 2010.

Hopkins, Brian, and Robin Wilson. "The Truth About Konigsberg." *College Mathematics Journal* 35, no. 3 (2004).

Humphreys, Lisa, and Ray Shammas. "Finding Unpredictable Behavior in a Simple Ordinary Differential Equation." *The College Mathematics Journal* 31, no. 5 (2000).

Peterson, Ivars. "MathTrek: Rock-and-Roll Bridge." *Science News*, January 30, 1999. http://www.sciencenews.org/pages/sn_arc99/1_30_99/mathland.htm.

Picon, A. "Navier and the Introduction of Suspension Bridges in France." *Construction History* 4(1988).

Simone Gyorfi

The Mathematics of Bridges

A bridge has to support various forms of forces: tension, compression, bending, torsion, and shear. It has to carry its own weight (or "dead load"), the weight of the traffic for which it was intended (or "dynamic load"), and it should resist various natural forces, such as wind or earthquakes. The Tacoma Narrows Bridge is often presented in engineering, physics, or mathematics classes as an application of oscillation problems or differential equations. It was a 1.1 mile (1.9 kilometer) long suspension bridge and collapsed in 1940—four months after being opened—because a 35–46 mile per hour wind produced an oscillation, which ultimately broke the entire construction.

Calculators in Society

Category: Communication and Computers.
Fields of Study: Algebra; Measurement; Number and Operations.
Summary: Advancements in calculator technology have profoundly changed society and mathematics education.

In the decades since the invention of a truly handheld calculator, these devices have evolved from four-function curiosities costing hundreds of dollars to sophisticated machines capable of performing a wide range of mathematical and statistical functions at the same cost as that "four-banger" from the early 1970s. The effect on society has been considerable, as the laborious arithmetic involved in routine scientific or financial calculations can be done by nearly anyone with minimal effort and accuracy that was unthinkable in the 1950s. A variety of technological advances and a new market for calculating power during the 1970s led to the "calculator wars" among a variety of manufacturers, and frequent major advances in the power of a calculator were marketed to a willing society. These powerful calculators have changed the school mathematics curriculum in a variety of ways and brought a new focus to the Advanced Placement (AP) calculus exams.

Early History of Calculators

In their 1951 textbook *Mathematics of Investment*, Paul Rider and Carl Fischer made occasional reference to the ability of "computing machines" to facilitate involved calculations in financial mathematics. However, since such machines were by no means common in the 1950s, the book includes 123 pages of numerical tables, roughly one-third of the book's total length. These references were essential to actuarial calculations for many years, and their analogous tables of values of trigonometric functions, exponentials, and logarithms were a staple of mathematics textbooks for a comparable time period.

The rapid rise of low-cost electronic calculators—a generation beyond the electric computing machines to which Rider and Fischer referred—reduced those tables to a mere historical curiosity within a generation. In 1958, Texas Instruments (TI) engineer Jack Kilby invented the integrated circuit, which became known as the "calculator on a chip," that revolutionized the world of calculating devices. Large electromechanical desktop calculators soon gave way to more compact electronic machines, which culminated with the development of the Cal-Tech handheld calculator in 1965 at Texas Instruments. The Cal-Tech was a simple four-function calculator that used a paper tape for output. With a new standard for what was possible, the rush to advance calculating machines, both handheld and desktop, was on.

Engineers at Hewlett-Packard (HP) merged the old with the new in 1968 with the development of the HP-9100A, the first fully electronic desktop calculator. The 9100A was considerably larger than the Cal-Tech but was much more versatile. Its function set included all of the functions found on a modern scientific calculator—trigonometric functions, logarithms, reciprocals, and others—and it was fully programmable. On viewing the 9100A, company founder Bill Hewlett included among his words of praise for the developers the challenge that the world needed a similar machine that would fit into a shirt pocket.

In 1972, Texas Instruments introduced the Datamath, a four-function calculator released under the TI name. This was a departure for the company, which until then had confined its calculator work to manufacturing parts for other companies' machines. Indeed, the Cal-Tech was built primarily to show other manufacturers what the company's parts could do, not as an eventual consumer product. In that same year, Hewlett-Packard engineers developed the HP-35, a fully scientific calculator that could fit into a shirt-pocket. With these two companies at the forefront of a rapidly advancing technology, and with many other manufacturers in close competition, the "calculator wars" began. The rapid evolution of affordable competing calculators from a variety of manufacturers went on throughout the 1970s and into the early 1980s.

A major innovation was TI's introduction of the TI-30 scientific calculator, which sold for under $30 beginning in 1976. The full scientific function set of the TI-30 on a low-priced machine was a huge advance over the $395 price tag of the original HP-35, and the TI-30 was regarded for many years as the best-selling calculator of all time.

HP introduced the first handheld programmable calculator, the HP-65, in 1974 (fewer than two years after its first scientific calculator), and followed it up in 1977 with the HP-67. TI countered with the SR-52 in 1975, which was succeeded by the TI-58 and TI-59 in 1977. Each of these milestone calculators allowed the user to specify a sequence of steps into a special memory. These steps could then be repeatedly executed as many times as desired. The HP models and the 52 and 59 provided the option of recording programs onto small magnetic cards for permanent storage, while the 58 and 59 came equipped with a slot for read-only memory (ROM) cartridges with space for dozens of specialized prewritten programs that were stored on the chip and could be run as needed without the need for repeated keying.

Special-Purpose Calculators

Special-purpose calculators are preprogrammed with functions and formulas that are specific to a particular profession or interest. Among the earliest were calculators designed for financial mathematics, with keys and routines for solving the time value of money problems and automating interest calculations—and here was where Rider and Fischer's prediction was exceeded. These business calculators were considerably more sophisticated than could have been imagined in 1951.

By far the most successful business calculator is Hewlett-Packard's HP-12C, which was introduced in 1981 and is still in production 30 years later. In most senses, the 12C is the industry standard financial calculator, and it has been the key to HP's successful focus on the business calculator market. In 2003, the 12C got a facelift—and a faster processor—as the HP 12C Platinum Edition.

Unit conversion calculators inspired by the push in the 1970s to introduce the metric system in the United States live on in a variety of construction calculators, many of which have been produced by a small company, Calculated Industries (CI). CI was founded in the 1978, and its first product was a real estate calculator dubbed "The Loan Arranger." Future financial calculators from CI would expand in capability to accommodate more sophisticated calculations, and a separate line of CI financial calculators is specific to Canadian interest calculations. Later product lines from CI included the Construction Master and Measure Master lines—which were specialized for the building industry. CI also produces a series of electrical engineering calculators and a pair of professional plumbing calculators.

CI also manufactures special-purpose calculators for a variety of niche markets. Do-it-yourselfers can find the calculations they need preprogrammed into the ProjectCalc series. Several of these have been rebranded by Sears under the Craftsman line. The KitchenCalc Pro is preset to convert cooking measurements and includes a built-in timer. The Quilters' FabriCalc is one of the company's most successful hobbyist calculators and automates the considerable mathematics involved in quilting. Most recently, the Mr. Gasket Hot Rod Calc was developed to serve performance automotive enthusiasts with a collection of functions for use in assessing an automobile's performance.

Calculators in the Classroom

In 1976, Texas Instruments released the Abstract Linking Electronically (ABLE) calculator system, which represented the first attempt to manufacture a calculator specifically designed for elementary school classrooms beginning in the earliest grades. The ABLE system consisted of a standard four-function calculator with six interchangeable faceplates. These faceplates blocked access to some of the calculator's functions and could be switched out to allow a richer selection of options as a child's mathematical sophistication grew.

There was then, and continues to be, considerable tension over the question of calculator use in school mathematics. The conflict is generated by the ability of inexpensive calculators to automate routine arithmetic problems, which had led one side of the debate to suggest that there is no need to require computational automaticity, such as memorizing multiplication tables, which a calculator can handle. These advocates then assert that calculators free up room in the curriculum for what are called "higher-order" mathematical thinking skills. Those opposed to this view assert that higher-order skills are not useful without a sound foundation based on mastery of routine calculations. Sensible middle ground exists between these two viewpoints, and a variety of combinations of these approaches are advocated in textbooks and available to teachers.

In 2000, TI expanded the Explorer line to include the TI-15 Explorer calculator, which was designed for use in grades 3–6. This calculator contains specialized keys for computations like place value calculations and fraction operations without cluttering the keyboard with higher-level computations, like trigonometric functions, that are not studied in elementary school. Additionally, the TI-15 Explorer includes two keys that can be programmed to repeat simple operations, a randomized arithmetic tutor, and tools for exploring inequalities. A simpler companion calculator, the TI-10, was introduced in 2002 and is aimed at kindergarten through third grade classrooms.

At higher grade levels, one effect was far less controversial. With the advent of inexpensive powerful scientific calculators, there was no longer a need for extensive tables of functions in precalculus textbooks.

Graphing Calculators

In 1985, Casio introduced the first graphing calculator, the fx-7000G. In addition to serving as a fully functional scientific calculator, the fx-7000G had a large (1.4-by-2-inch) LCD screen on which graphs of functions could be displayed. This allowed students to work with functions from both numerical and graphical perspectives, and set the stage for a revolution in mathematics teaching. Graphing calculators soon came to be seen as one of the primary components of this shift in teaching and learning.

Hewlett-Packard advanced handheld capacity further with the HP-28C, introduced in 1987. In addition to numerical and graphical approaches to functions, the 28C was able to perform symbolic algebra and calculus, working with variables directly without the need for numbers. Texas Instruments released its first graphing calculator, the TI-81, in 1990, and the TI-85 soon after. The TI-82, 83, 84+, 86, and 89 have extended this successful product line, while the TI-80 and 73 have reached downward into middle schools.

As graphing calculators and computer algebra systems, such as Derive and Mathematica, competed for space in calculus classrooms around the world, it became clear that standardized testing would have to accommodate these new devices. Beginning in 1995, the Advanced Placement calculus exams have required the use of a graphing calculator on part of the exam, one that can plot graphs of functions, solve equations numerically, compute numerical derivatives, and evaluate definite integrals numerically. The College Board, which administers the AP exams, draws the line at calculators with a typewriter-style QWERTY keyboard, such as the TI-92 (introduced in 1996) and Voyage 200 (introduced in 2002) from Texas Instruments. The concern here is for the security of the tests, as the typewriter keyboard and text-processing capability are thought to make it too easy to collect confidential test questions and remove them from the testing site.

The Future of Calculators

It is unclear what new ground remains to be broken in future calculators. Three-dimensional graphing is available on a variety of TI and HP machines, but the size of the screen and the challenge from computer algebra systems, such as Mathematica, have limited the reach of this feature. Calculating power is finding its way into a variety of other handheld devices. Just as many people no longer wear watches because they can get the time from their cell phones, calculator applications for cell phone platforms may render the cell phone an attractive alternative to a specialized calculator. While there are cost and durability issues to be considered in this comparison, CI has recognized this alternate platform by marketing its Construction Master Pro software for the iPhone.

Further Reading

Ball, Guy, and Bruce Flamm, *The Complete Collector's Guide to Pocket Calculators*. Tustin, CA: Wilson/Barnett Publishing, 1997.

Hicks, David G. "The Museum of HP Calculators." http://www.hpmuseum.org.

Sippl, Charles J., and Roger J. Sippl. *Programmable Calculators*. Champaign, IL: Matrix Publishers, 1978.

Woerner, Joerg. "Datamath Calculator Museum." http://www.datamath.org.

Mark Bollman

Calendars

Category: Space, Time, and Distance.
Fields of Study: Measurement; Number and Operations; Representations.
Summary: Various calendars use different methods of resolving the need for "leap" days, months, or years.

Even the earliest human beings must have noticed the astronomical cycles: the alternation of day and night, the pattern of the changes in the moon's shape and position, and the cycle of the seasons through the solar year. It must have been frightening every autumn as the days became shorter, causing concern that the night might become permanent. This led to celebrations of light in many areas as the days began to lengthen again. Once the repetitions of the patterns were recognized, people could count them to keep track of time. Longer cycles helped avoid difficulties in keeping track of large numbers—once approximately 30 days had been counted, people could, instead, start counting "moons." This same technique of grouping also occurred in the development of counting systems in general—leading to place-value structures in numeration systems.

The problem was that the shorter cycles did not fit evenly into the longer cycles. Trying to fit the awkward-length cycles together actually led to some mathematical developments: two different cycles would come together at the least common multiple of the lengths of their cycles; modular arithmetic and linear congruences were methods of handling leftover periods beyond the regular cycle periods.

The Julian and Gregorian Calendars

The Romans developed the Julian calendar (named for Julius Caesar), recognizing that the exact number of 365 days in one year was slightly too short and would soon throw the calendar off the actual cycle of the solar year. They found a remedy by assuming the solar year to be 365.25 days. To handle the one-quarter day, they added one full day every four years—the day that we call "leap-year day" on February 29 of years whose number is a multiple of four. This gives $3(365) + 366 = 1461$ days in four years, or an average of 365.25 days per year as desired. However, the actual solar year is 365.2422 days long (to four decimal places), about 11 minutes less than the Romans' value. Even in a human lifetime, this is negligible. Over centuries, however, the extra time builds up so that by the 1500s, the calendar was 10 days off from the solar cycle (for example, the vernal equinox seemed to be coming too late).

In 1582, Pope Gregory XIII assembled a group of scholars who devised a new system to fit better. It kept the Roman pattern except that century years (1600, 1700), which should have been leap years in the Roman calendar, would not have a February 29 unless they were multiples of 400. For example, 1900 was not a leap, year but 2000 was. In the full 400-year cycle, there are (400 × 365) regular days + 97 leap-year days = 146,097 days, making an average of 365.2425 days per year. This cycle is only .0003 days (about 26 seconds) too much; in 10,000 years, we would gain three extra days. This system was called the Gregorian calendar. Since the longer Julian calendar had fallen behind the solar year by about 10 days, the changeover to the Gregorian required jumping 10 days.

Various countries in Europe changed at different times, with each switch causing local controversy as people felt they were being "cheated" out of the skipped days. The effects of the change are noticed in history. When Isaac Newton was born, the calendar said it was December 25, 1642; but later England changed the calendar, so some historians today give Newton's birthday as January 4, 1643. The Russians did not change their calendar until after the 1917 October Revolution, which happened in November by the Gregorian calendar.

The Lunar Calendar

The other incongruity of calendar systems is that the moon cycle of 29.53 days does not fit neatly in the 365.2422 days of the year. Twelve moon periods is 11 days shorter than a year, and 13 "moons" is 18 days too long. It is interesting to note that of the three major religious groups of the Middle East—the Christians, the Muslims, and the Jews—each chose a different way to handle "moons/months." The Christians (actually, originally, the Romans) ignored the moon cycle and simply created months of 30 and 31 (and 28 or 29) days. The Muslims considered their year to be 12 moon cycles and ignored the solar year. This means that dates

A 1412–1416 illumination depicting the month of March with the constellations of the zodiac on top. (Photos.com)

in the Muslim calendar are shifted back approximately 11 days each year from the solar calendar, and Muslim festivals move backward through the seasons.

People in the Jewish faith chose to keep both the solar and lunar cycles. After 12 lunar months, a new year begins—as in the Muslim calendar—11 days "too early." However, after the calendar slips for two or three years—falling behind the solar calendar by 22 or 33 days—an extra month is inserted to compensate for the loss. There is a 19-year pattern of the insertion of extra months, which keeps the year aligned with the solar year. Interestingly, the traditional east Asian calendar follows a pattern very similar to the Jewish calendar.

The Mayan Calendar

The Mayans of Central America had a very complex pattern of cycles leading to a 260-day year for religious purposes, and a regular solar year that was used for farming and other climate-related activities. Their base-20 numeration system, which should have had place-value columns of 1-20-400-8000, was adjusted to 1-20-360 to fit into the 365+ days of the year. They were also notable for developing massive cycles of years lasting several millennia, including one ending in late 2012 of the Gregorian calendar.

Further Reading

Aslaksen, Helmer. "The Mathematics of the Chinese Calendar." http://www.math.nus.edu.sg/aslaksen/calendar/chinese.shtml.

Crescent Moon Visibility and the Islamic Calendar. http://aa.usno.navy.mil/faq/docs/islamic.php.

Duncan, David Ewing. *Calendar: Humanity's Epic Struggle to Determine a True and Accurate Year*. New York: Harper Perennial, 2001.

Rich, Tracey R. "Judaism 101: Jewish Calendar." http://www.jewfaq.org/calendar.htm.

Richards, E. G. *Mapping Time: The Calendar and Its History*. New York: Oxford University Press, 2000.

Stray, Geoff. *The Mayan and Other Ancient Calendars*. New York: Walker & Company, 2007.

LAWRENCE H. SHIRLEY

Canals

Category: Architecture and Engineering.
Fields of Study: Algebra; Geometry; Number and Operations; Problem Solving.
Summary: Modern canal design, particularly the challenges of a lock system, depends on partial differential equations and other mathematics.

Canals are human-made channels for water, including both waterways big enough to be traversed by ship (built for transportation), and aqueducts (built for water supply and irrigation). The building of canals was critical to the formation of many ancient civilizations, which needed to manipulate water access in order to enable an early urban lifestyle. Many ancient mathematics texts address such large-scale ancient engineering projects.

A number of the surviving Babylonian tablets dealing with geometry were composed for canal projects: they calculated the number of workers necessary to build the canal in a given number of days, the dimensions of the canal, and the total wage expenses so that the ruler for whom they were built would know how much the project would cost. Mathematical problems related to the construction of canals can also be found in the fifth chapter of *The Jiuzhang Suanshu* (*Nine Chapters on the Mathematical Art*), one of the earliest surviving ancient Chinese mathematics texts. Mathematicians and engineers have long investigated canals.

For instance, Jacopo Riccati worked on hydraulics and constructed dikes in Venice, and Barnabé Brisson employed descriptive geometry in the design and construction of ship canals. Mathematicians like George Green and Joseph Boussinesq analyzed and modeled wave motion in canals. John Russell tested and studied steam-powered canal transportation and wave creation for the Union Canal Company. Mikhail Lavrentev created a theoretical foundation for large projects on the Volga, Dnieper, and Don rivers. Mathematics theories and techniques are critical when engineers, mathematicians, and software programmers model the changing flow rates and levels of a canal. They rely on mathematics like the Saint-Venant equations (partial differential equations that are named after mechanic and mathematician Jean Claude Saint-Venant).

The simplest canals are merely trenches through which water runs, usually lined with some kind of construction material. Canals need to be level in order to be navigable (a ship cannot move "uphill"). When the land itself is not level, a lock system must be used. Locks are systems for raising and lowering boats from one stretch of water to a stretch of water at a different level. The most common type of canal lock—used in ancient China, and most likely in the ancient West, and still common today—is the pound lock, which consists of a watertight chamber with gates at either end to control the water level in the chamber.

Engineer Chiao Wei-Yo is credited with the design of the lock system, which he used on the Grand Canal in the tenth century. In the pound lock system, a ship enters the chamber (the "pound") from one length of canal; water is raised or lowered to bring the ship to the level of the next length of canal; and the ship exits the chamber. The necessity of locks added much complexity, time, and room for error to the construction of canals, which would have been sufficient to discourage Napoleon's aims. In 2010, the Panama Canal commemorated its one-millionth transit, and engineers plan to expand the canal by adding more locks. It has been referred to as one of the seven wonders of the industrial world.

Further Reading
Bernstein, Peter. *Wedding of the Waters: The Erie Canal and the Making of a Great Nation*. New York: W. W. Norton, 2006.
Karabell, Zachary. *Parting the Desert: The Creation of the Suez Canal*. New York: Alfred A. Knopf, 2003.
Montañés, Jose. "Mathematical Models in Canals." In *Hydraulic Canals: Design, Construction, Regulation and Maintenance*. New York: Taylor & Francis, 2006.
Parker, Matthew. *Panama Fever: The Epic Story of One of the Greatest Human Achievements of All Time*. New York: Doubleday, 2007.

Bill Kte'pi

Major Canals

Significant canals include the Erie Canal in the United States, the Suez Canal in Egypt, the Panama Canal in Panama, and the Grand Canal in China, each of which was constructed as a major operation for the sake of hastening trade and transport. Judge Benjamin Wright, who some call the father of American civil engineering, was appointed the chief engineer of the Erie Canal. Astronomer and mathematician Guo Shoujing (also known as Kuo Shou-ching) was the head of the Water Works Bureau in the thirteenth century. He made improvements to control the water level in existing canals and built new ones.

The Suez Canal was imagined long before it was completed, and the Egyptians were masters of large-scale engineering projects. Napoleon Bonaparte, during the French invasion of Egypt, reportedly discovered ruins of an ancient canal, which inspired him to order a preliminary survey exploring the possibility of a north–south canal joining the Mediterranean and the Red Sea (the ancient canal had been east–west and was intended to link the Red Sea and the Nile). The project was abandoned—possibly because of the belief that the Red Sea was higher than the Mediterranean—and so the canal remained unbuilt for 70 years.

Carpentry

Category: Architecture and Engineering.
Fields of Study: Algebra; Geometry; Measurement.
Summary: Precise measurement is the foundation of the building trades.

While the word "carpentry" originally comes from the Latin root for chariot maker, today, the term refers to a number of trades that use wood for the construction of buildings and other articles. As there is a wide range of activities involved in carpentry tasks, carpenters must possess many different manual and intellectual skills to function in the profession.

Types of Carpenters

Carpenters who work on houses often fall into one of two broad categories: framing carpenters who work on the rough frame of a building, and finish carpenters who complete trim, stairs, railings, shelves, and other detail work. However, in practice, many carpenters end up doing some of each type of work, and carpenters who specialize in remodeling may not only do framing and finish carpentry but also tasks that are not strictly carpentry at all, such as plumbing, wiring, sheetrock finishing, and painting. There are also carpenters who specialize more narrowly, such as cabinet makers or carpenters who work on the specialized joinery between large posts and beams required in timber frame and log cabin construction.

Tasks of the Carpenter

Carpentry requires a variety of skills, including reading blueprints, measuring, cutting, fastening, and finishing. In addition, a carpenter must have knowledge of materials, including a variety of wood products and fasteners; and tools, including measuring devices, saws, drills, hammers, planes, and sanders. Carpenters who work on their own or as subcontractors on larger jobs must also have skills in cost-estimation and billing.

Consider, for instance, a carpenter who has been hired to add a covered deck onto a house. This carpenter might begin by working with the homeowner to determine the size and shape of the deck, possibly using a Computer Assisted Design (CAD) program to generate three-dimensional representations of how the finished project will look. After deciding on a design, the carpenter will need to use structural engineering tables to assess structural issues related to the design, such as the dimensions required for posts, the placement and size of cross-bracing, and the sizes of timbers that will be needed to span the distance between posts. From the calculations, the carpenter will then generate a price estimate, based on a materials list and an estimate of labor. The actual construction will include pouring concrete footers for the posts, measuring and cutting posts and joists with a circular saw, fastening materials to one another and to the house, screwing decking materials to the framing, framing a roof, installing roofing materials, constructing a railing, and building and finishing a set of stairs from the yard to the deck.

A Carpenter's Calculations

In the process of creating a simple covered deck, this carpenter will be making many measurements, calculations, and decisions regarding:

Carpenters need to be able to read blueprints, and measure, cut, fasten, and finish a variety of materials. (Photos.com)

Layout: The initial position of the deck must be laid out so it is square to the house. To do this, the carpenter will construct a set of batter boards that are set outside the corners of the proposed deck and allow strings to be pulled to mark the edges of the deck. Employing the rule that the diagonals in a rectangle are equal to one another, the carpenter adjusts the strings to bring the corners to 90 degrees. Corner square may also be established and checked using the Pythagorean theorem.

Footers: Each post will be anchored to a concrete footer that will prevent it from moving or sinking into the ground. The bottom of the holes for these footers must be dug below the freeze level for the geographic area where the deck is being built so that the footers will not be heaved out of place by the freezing and resulting expansion of the soil. By consulting the building code, the carpenter will determine the appropriate area for the footer in square feet; multiplying by the height will give the cubic feet. If this is a large project, where the concrete will be delivered, the carpenter will have to convert cubic feet to cubic yards, as this is the unit in which concrete is ordered.

Raising the Posts: After pouring the footers, the carpenter will raise the posts for the deck being built. Since these posts will also support the roof in this example, they must be cut carefully to take into account any variation in the height of the footers. This measurement will be done by using a transit, a laser level, or a water level to assess the difference in the height of the footers. The carpenter will then add or subtract length to the height of each post to compensate. Once the posts are cut, they can be raised into position, ensuring each is plumb (perfectly vertical) using a level.

Joists and Decking: The sizing for all the wooden parts of the deck is determined by calculating how long a distance must be spanned and the weight the span will carry. The timber that is parallel to the house and runs between the posts must be sized to be strong enough to carry all the weight between each pair of posts; the longer the span between posts, the larger this timber must be. Similarly, the floor joists that butt into this timber will need to be large enough to carry the weight over their length, and the decking will be sized so that it does not sag between the joists.

Fasteners: In our example, the deck will be fastened to the building using bolts, and held together using nails, while the decking itself will be screwed on. The carpenter has many fasteners to choose from with many different finishes. Each type of fastener has special characteristics that make it useful for certain tasks. Nails are typically sold by the pound and come in sizes from large 20d framing nails (often called 20 penny nails) to small 6d finish nails. Screws are also sold by the pound but are sized by length and by a number that can be converted, using a chart, to their diameter. Bolts are sold by diameter and length; as is the case with all fasteners, there are many different types among them, lag bolts, carriage bolts, and through bolts.

The Roof: The roof over the deck will be set at an angle so water runs off it and away from the house. The pitch of a roof is typically measured in "rise over run," with the denominator of this fraction always given as 12. Thus, a roof that goes up four feet over a run of 12 feet is said to be a "4:12 roof." The carpenter will use a special tool called a "speed square" that allows the direct conversion of roof pitch to angles and mark rafters for cutting.

Stairs: While stairs can be constructed to be more or less steep, a carpenter must keep in mind a basic mathematical relationship between tread length and riser height that will make a set of stairs comfortable to ascend. It turns out that because of the characteristic of the human gait, the steeper a stair, the less wide each tread should be. The formula that carpenters use is that for each stair, twice the rise plus the run should equal 24–26 inches.

Of course, once the carpenter is done with the project, there are still numerous other tasks to complete, including building railings and benches, as well as finishing and waterproofing the surfaces. If the homeowner were to want an outdoor grill area, with built-in cabinets, the carpenter would have a whole new set of challenges worthy of a cabinet maker and finish carpenter.

Further Reading

Gerhart, James. *Mastering Math for the Buildings Trades*. New York: McGraw-Hill, 2000.

Webster, Alfred P. *Mathematics for Carpentry and the Construction Trades*. 2nd ed. Upper Saddle River, NJ: Prentice Hall, 2001.

Jeff Goodman

Castles

Category: Architecture and Engineering.
Fields of Study: Algebra; Geometry; Measurement.
Summary: Mathematics has been used to both construct and study castles.

Castle are fortified structures, used as residences by European nobles in the Middle Ages. Early castles were often made of wood, but with the development of better attack methods, castle builders switched to stone as the main building material. With the extensive use of artillery, residential castles became indefensible. They were replaced by purely military forts (not used for administrative and residential purposes) and decorative residences resembling castles (not used during wars). The geometry of a castle was often dictated by defense considerations. Architect Benjamin Bramer fortified castles and published a work on the calculation of sines.

The Alhambra, a fourteenth-century palace and fortress, is well known for its mathematical tiles. In the early twenty-first century, the American Institute of Mathematics proposed a headquarters in California that would be modeled after the Alhambra, popularly referred to as a "math castle."

Castles are frequently found in fantasy and horror literature. One common image is that of Dracula's castle. Dracula author Bram Stoker earned a degree in mathematics. Some mathematics teachers use castles like Cinderella's castle or sand castles to explore concepts such as ratios, fractions, volume, statistics, and geometric shapes. Scientists, including physicist Mario Scheel, explore the physical properties of sand-like material, and researchers in experimental archaeology model and design castles.

Geometry of Castle Defense

Both the layouts of castles and the shapes of their parts were dictated by defense needs. For example, concentric castles consisted of several concentric walls. The barbican (the outer wall) had relatively many entrances, while the inner wall had few, making the attacking army crowd between walls and, thus, become vulnerable to defenders.

Keeps and towers were mostly round to allow for a larger arc of shooting coverage from each arrowslit. In addition, the isoperimetric theorem states that for a given area, the circle has the least perimeter among all shapes, thus minimizing the amount vulnerable walls (not to mention reducing the costs of building materials). Each corner introduced blind spots where enemies could avoid arrows, and circles have no corners. Also, corners are more vulnerable for mining.

Cylindrical towers led to the invention of spiral staircases. Most castle staircases were built so attackers would ascend clockwise, making the central shaft of the staircase interfere with their right hands—often the hand that held the sword.

Stonemasons building castles used simple tools, such as compasses, dividers, and straightedges. Their manuals included descriptions for creating a variety of shapes with these tools. For example, pointed and rounded arches, including Tudor, lancet, and horseshoe arches, could be traced with compasses and straightedges.

Shooting from high towers allowed for better view, and also used gravity to add acceleration to arrows and other projectiles. When glass windows were installed in circular towers, they were made by blowing glass inside a cylinder, cutting it, and then connecting multiple pieces with lead to match the curvature of the castle wall.

Castle builders used terrain geometry to support defense. In addition to the height advantage of the castle walls and towers, castles were frequently situated on hills (either natural or artificial) or on earthen mounds called "mottes." Defensive ditches around castles, called "moats," prevented siege towers from coming close. When moats were filled with water, they could also make digging tunnels for mining the walls more difficult.

The construction of moats led to the invention of drawbridges and the mechanisms of raising and lowering them. The drawbridge mechanisms involved levers and pulleys.

Logistics and Finance

Building a large castle was a major financial undertaking spanning many years, and occasionally bankrupted the ruler attempting it, such as King Edward I. Supplying the castle, especially with enough supplies to withstand lengthy sieges, presented another organizational problem. A siege was a common method of castle attack in which the attackers surround the castle grounds and waited for the defendants to starve. The siege process could sometimes last for months or even years.

Experimental archaeology is a new field of study that combines archaeological research, computer modeling, and actual building. Observations in building experiments allow for conclusive results of how models can be made to work. For example, Project Gueledon is a real-size castle built recently to help give people a deeper understanding of how castles were constructed in medieval times. The researchers used building methods and materials similar to those used by thirteenth-century castle builders, with a team of 50 workers from various professions.

Further Reading

Holden, Constance. "A Castle Fit for a Mathematician." *Science* 314, no. 6 (October 2006).

Whitney, Elspeth. *Medieval Science and Technology*. Westport, CT: Greenwood Press, 2004.

Maria Droujkova

Cell Phone Networks

Category: Communication and Computers.
Fields of Study: Algebra; Geometry; Measurement.
Summary: Mathematics is involved in the design of the cell network and the assignment of calls to frequencies, as well as in data compression and error compression.

Cell phones have grown from a novelty, to a luxury, to a virtual necessity since the 1990s, with the number of cell phone subscribers in the United States growing from about 91,000 in 1985 to 276 million in 2009. Part of the reason that cell phones have become so reliable, cheap, and secure has to do with mathematics. Mathematics is involved in the design of the cell network and the assignment of calls to frequencies (or channels), as well as data compression and error compression that allow a large number of clear calls to be carried over a small bandwidth. The concept of a tree from graph theory can be used to understand cell phone networks, which are challenging because of the large amount of data and links. Mathematicians like Vincent Blondel analyze millions of users and months of communication.

Cellular Radio Networks

Cell phones work by communicating via radio signal with a nearby cell phone tower. In a cellular radio network, the type of system used for cell phone coverage, the land area to be supplied with coverage is divided into regular shaped regions (or "cells"), each of which has a corresponding radio base station or cell tower. Phones within a particular cell connect via radio signal to the tower for that cell, which then connects to the public telephone network through a switch. The range of a tower may be about one-half mile in urban areas up to about five miles in flat rural areas.

Because of this relatively short transmission range, cell phones and towers can use low power transmitters. In addition to allowing phones to be small and use smaller batteries, the low power also means the radio frequencies can be reused by towers not too far away from each other without any interference between the transmissions. This function allows cell phone networks to carry a larger number of calls in a smaller bandwidth. Typically, cell companies will divide their coverage area into regularly shaped cells or regions with each one covered by a single tower. In fairly flat areas, these regions are usually hexagonal in shape—an idea developed by Bell Labs engineers W. Rae Young and Douglas Ring in the middle of the twentieth century.

The frequencies used by a particular tower for transmissions in its region cannot be used by any of the six regions with which it shares a boundary. The Four Color Theorem from graph theory indicates that only four frequencies are needed to ensure that regions that share a boundary do not use the same frequency. However, companies usually want to further buffer the distance between reuse of the same frequency, so they divide the frequencies up into seven bundles and use a different one on each of the six cells sharing a boundary with a given cell.

Cell Phone Channels

During the twentieth century, there were many discussions among professionals at the Federal Communications Commission regarding the possibility of opening up frequencies for phone use. Cellular networks began to appear around the world. For instance, Japan offered a 1G system in 1979, and, in 1983, AT&T and Ameritech tested a commercial cellular system in Chicago. Much of the advancement in cell network technology has been focused on the frequency band within a cell, which must be divided up to carry several calls at the same time. In first-generation cell technology, calls were transmitted in analog, which allowed only one call per frequency. Typically, a cell phone carrier was assigned 832 radio frequencies to use in a city. Each call was full duplex, meaning that it used two frequencies: one to transmit and one to receive.

Thus, typically there were 390 voice channels with the remaining 42 radio frequencies used for control channels that were used to locate and communicate with phones but not to carry calls. If the 395 voice channels were divided into seven frequency bundles, that made 56 voice channels per region. So if more than 56 calls were in progress in a given region at a given time, then one of the calls would be disconnected or dropped. Fortunately, first-generation technology is no longer in use. With second-generation (2G) cell technology, calls were no longer analog signals but were converted to a digital (0 and 1) format. This shift is similar to the change from cassette tapes to compact discs in the recording industry.

The greatest advantage to digital technology is that it allows for sophisticated data compression techniques to be used without losing acceptable call quality. Data compression allows for between three and 10 digital calls to be carried in the bandwidth necessary for a single analog call. Further advancements in compression have allowed for even newer third-generation (3G) technology. 3G networks have much faster transmission speeds and allow the use of smartphones that can transmit data fast enough to surf the Internet, send and receive e-mail, and even instant message with a cell phone. Newer 4G technology adds even more speed and capacity to cell phone networks.

Further Reading

Agar, Jon. *Constant Touch: A Global History of the Mobile Phone.* Cambridge, UK: Icon Publishers, 2003.

Brain, Marshall, Jeff Tyson, and Julia Layton. "How Cell Phones Work." *HowStuffWorks.* November 14, 2000. http://electronics.howstuffworks.com/cell-phone.htm.

Mark Ginn

City Planning

Category: Architecture and Engineering.
Fields of Study: Geometry; Number and Operations; Problem Solving.
Summary: Mathematics is used to model optimal city designs and reduce problems of traffic congestion, sanitation, and water distribution.

Also called "urban planning" and "town planning," city planning is a discipline that focuses on the various economic, environmental, historical, physical, political, and social characteristics of the urban environment and their harmonious organization. It encompasses a variety of projects, processes, and goals that involve multiple disciplines and fields of expertise, such as physical design, and quantitative and qualitative research, as well as analysis, forecasting, strategic planning, negotiation, and public mediation. Since the late nineteenth century—and especially during the second half of the twentieth century—the profession has increased its reliance on statistics and mathematics.

Early History

The early origins of urban planning can be traced in the physical design and purposeful spatial organization of some ancient cities in Mesopotamia, Egypt, the Mediterranean Basin, South and Central America, the Yellow River Basin, and along the Indus Valley. Many of these settlements present a hierarchical system of paved streets, often following a rectilinear grid, with water supply and drainage systems. The Middle Ages was not a propitious era for urban planning. It became popular again during the Italian Renaissance with the design of ideal cities. Influenced by the belief that a perfect form was the image of a perfect society, designers opted for radial or centrally planned cities frequently uniting the perfect geometric figures of the square and circle into a star-shape layout. In the seventeenth century, the rise of nation states and absolutism was conducive to the development of the monumental baroque city with its straight and endless avenues, unbroken horizontal rooflines, and repetition of uniform elements, which glorified the ruling power. Simultaneously, the advances of warfare techniques led to the disappearance of the old city walls and the adoption of new complicated systems of fortification with considerable outworks and bastions in spearhead forms.

The Industrial Era

The modern origins of city planning have their roots in the industrial city of the mid- and late nineteenth century. In both Europe and the United States, rapid technical progress, tremendous industrial development, and massive displacements of rural population to urban areas created considerable problems that threatened to disrupt the existing social order. The dreadful conditions experienced by masses of people living in abject poverty and misery in overcrowded slums sprawling around wealthier districts became a source of concern for the general public health. In 1854, Dr. John Snow—the father of modern epidemiology—identified the source of a cholera outbreak in London by studying the patterns of the disease and using statistics and a spot map illustrating the clustered death cases of cholera around the Broad Street pump. The fears of major epidemics resulted in the rise of a social movement for urban reform and planning, which first focused on water supply and sanitation improvement, and later on housing provision.

In the 1880s, the basic lack of information regarding the extent and distribution of poverty in London led English philanthropist Charles Booth to develop a comprehensive and scientific social survey investigating the incidence of pauperism first in East London, and later in the entire city. His quantitative statistical analyses and qualitative research presented in 17 volumes with accompanying colored maps indicating the levels of poverty and wealth by street received considerable attention. They were also influential in demonstrating the importance of social surveys for public policy, demographics, and sociology as well as in improving census data collection.

Similar problems affected Paris and, after visiting London, Napoléon III placed considerable emphasis on urban planning to modernize the medieval capital into the capital of light. The large-scale restructuring program under the direction of Baron Haussmann affected not only the center of Paris but also its surrounding suburbs. At the time, it was the largest urban renewal project ever implemented. The plan created a network of large, easily accessible avenues and boulevards with radiating vistas terminated by prestigious public edifices and monuments. In addition to the building of 71 miles of new roads, the layout of 400 miles of pavement, and the doubling of the number of trees lining the streets, the city's infrastructure was entirely renovated. The construction of more than 340 miles of sewers and hundreds of miles of aqueducts increased the water supply by 400%. Haussmann also created two major urban parks and two large natural preserves on the periphery. This urban metamorphosis influenced the design of numerous cities worldwide and in particular the "White City" of the World's Columbian Exposition of 1893 in Chicago, which was the first large-scale project of the City Beautiful movement in the United States. The aim of expanding civic consciousness and raising the standards of civic design

Lithograph by Currier and Ives of the 1893 World's Columbian Exposition in Chicago. The "White City" in the exhibition was the first large-scale planning execution of the City Beautiful movement in the United States. (Library of Congress)

culminated in the publication of the famous Plan of Chicago in 1909, which coincided with the first university course in city planning at Harvard and the first National Conference on City Planning. In 1917, the American City Planning Institute was founded.

The Twentieth Century

Nevertheless, with the growth of the automobile as a favorite mode of transportation, it was the "Garden City" concept invented by Ebenezer Howard that became the leading model for the development of U.S. suburban residential communities. As the old city centers became increasingly congested, transportation planning became increasingly important to ensure an efficient balance between land-use activities and the potential communications between them. Transportation planners regularly collect data, which they analyze and process, to forecast future traffic using various techniques such as land-use ratio methods, multiple regression models, category analysis, growth-factor methods, synthetic models, modal split analysis, diversion curves, and geographic information systems (GIS).

After the U.S. Department of Commerce published "A Standard State Zoning Enabling Act" and "A Standard City Planning Enabling Act" in the 1920s and the U.S. Supreme Court upheld the constitutionality of zoning in 1926, most U.S. cities established planning departments to adopt master plans and zoning regulations that allowed them to control land-use development, protect property values, and segregate uses. Cities also started implementing subdivision controls and regulations. These new tools contributed to the belief in part of the planning community of the possible rational and scientific management of cities. On the other hand, idealists such as Frank Lloyd Wright and Lewis Mumford criticized the new pragmatic and technological approach, preferring a philosophy of city development for humanistic and social ends as epitomized in the design of Radburn, New Jersey. Over time, zoning regulations revealed some drawbacks. They often increased traffic congestion, and sometimes prevented the construction of affordable homes. Some courts struck them down as exclusionary.

City planning in the post–World War II era was dramatically affected by four significant federally funded programs: public housing, urban renewal, home mortgage insurance, and highway building. The miserable failure of urban renewal—and the urban crisis of the 1960s that ensued—required new approaches to urban planning. During the second half of the twentieth century, city planning became increasingly defined as a cyclical process attempting to balance conflicting social, economic, environmental, and aesthetic demands while implementing selected objectives and goals. Therefore, regular monitoring became necessary to test, evaluate, and review the strategies and policies adopted on a continuous basis. City planners regularly use a wide range of models ranging from basic descriptive statistics to more complex mathematical models that allow them to understand the nature of various urban components and forecast the consequence of change.

Because of the tremendous complexity of urban systems, models can provide only a simplified representation of the studied phenomena. Consequently, there is considerable attention and controversy regarding the choice of variables, and their level of aggregation and categorization, as well as the handling of time, specification, and calibration. Although deterministic models are the dominant type of predictive models used by urban planners, there has been some attempt at developing stochastic models. Urban planners are also concerned with the accuracy, validity, and constancy of the models they use. Most models tend to be topic specific, focusing, for example, on population, housing, employment, shopping, transport, or recreation, but integrated forecasting systems have become more common as there has been an increasing recognition of the interdependence of the various subfields of a city.

Further Reading

Field, Brian, and Bryan MacGregor. *Forecasting Techniques for Urban and Regional Planning.* Cheltenham, England: Nelson Thornes, 2000.

Freestone, Robert. *Urban Planning in a Changing World.* New York: Routledge, 2000.

Hall, Peter G. *Cities of Tomorrow: Intellectual History of Urban Planning and Design in the Twentieth Century.* 3rd ed. Malden, MA: Blackwell Publishing, 2004.

Kostof, Spiro. *The City Shaped: Urban Patterns and Meanings, Through History.* Boston: Bulfinch Press, 1993.

Krueckeberg, Donald A. *Introduction to Planning History in the United States.* New Brunswick, NJ: Center for Urban Policy Research, Rutgers University, 1983.

Lynch, Kevin. *Good City Form.* Cambridge, MA: MIT Press, 1984.

Moughtin, Cliff, et al. *Urban Design: Method and Techniques.* 2nd ed. Burlington, MA: Architectural Press, 2003.

Catherine C. Galley
Carl R. Seaquist

Clocks

Category: Space, Time, and Distance.
Fields of Study: Measurement; Number and Operations; Representations.
Summary: Clocks are devices for timekeeping and are used for a variety of mathematical calculations, including finding one's longitude.

The term "clock" in a generic sense is applicable to a broad range of devices for timekeeping usually concerning fractions of the natural unit of time—the day. Modern clocks operate through various physical processes. It does not matter what kind of periodic signals a clock produces—ringing a bell, firing a cannon, flashing a light, moving a hand, displaying a number, or generating electric impulses. Mathematics has been fundamental both in the design of clocks and in the measurement of their accuracy. Modular arithmetic, an algebraic concept involving cycles, is sometimes informally known as "clock arithmetic." In the realm of biology, mathematicians have also worked on theories related to the operation of humans' internal biological clocks and bacterial genetic clocks.

History of Clocks

In everyday English language, watches and other timepieces that can be carried individually sometimes continue to be distinguished from clocks. Via Dutch, Northern French, and Medieval Latin, the word "clock" is derived from the Celtic *clagan* and *clocca* meaning "bell." Those old clocks had a striking mechanism for announcing intervals of time acoustically. The history of clocks is much deeper, however. It started in early prehistoric times with sundials (often a vertical post or pillar on horizontal ground exposed to the sun or a post parallel to the Earth's axis) that were the first and oldest scientific instruments of archaic humankind. They worked only in the daytime. In the terminology of ancient Greece, such a device was called a *gnomon*, and the entire branch of science on sundials is *gnomonics*. Famous Egyptian obelisks—now reerected in some European capitals—were also sundials.

Timekeeping devices of different types were called *horologium* by the Romans. In its corrupted forms, this term later on entered many languages of the world. A noticeable step in the history of timekeeping was the invention of a "water clock" (the specific Greek name is the *clepsydra*). Water clocks could be used at night. Some of the water clocks in China and the Near East were quite large. Another type of simple clock was the "sandglass."

The modern era of clock-art started with the invention of weight-driven mechanical clocks (sometimes known as "chimes"). The inventor of such a novelty is unknown. Because daily prayer and work schedules in medieval times were strictly regulated, religious institutions required clocks, and it is certain that the earliest medieval European clockmakers were Christian clerics. Mechanical clocks were designed en masse in the thirteenth century in Western Europe. They were bulky and appeared on cathedral towers in many countries. Some of them have survived up to now and are among the great artifacts of the medieval epoch.

After the invention of tower clocks, efforts were made to design smaller pieces for tabletops and personal "pocket" clocks (watches) for individuals. Peter Henlein (c. 1480–1542), a locksmith from Nuremberg, Germany, is often credited as the forerunner of the first portable timekeeper, but this claim is disputed. His drum-shaped *Taschenuhr* was too big for a pocket. The first individual clocks were usually worn on the neck or beneath the knee. Timepieces of this type were often known as "Nuremberg eggs." The earliest clocks are very expensive now and are subjects for collectors.

Clocks for Navigation

A great chapter in clock-making began in conjunction with the rapid development of seafaring after the European discovery of the Americas. In order to determine one's position at sea, it is necessary to calculate two geographical coordinates: latitude and longitude. Latitude is easily computed directly from trivial astronomical considerations (the latitude of a locale is equal to the altitude of the celestial pole). As for longitude, it is equal to the difference between local time and the time of a prime meridian chosen specifically for cartographic purposes; navigators used different prime meridians in different countries in different epochs. To discover one's longitude, an observer must know the time at the prime meridian, which requires the art of "transporting" accurate time.

The search for accurate and convenient timekeeping became one of the most impressive scientific and technological challenges of the seventeenth century. Numerous mathematical and astronomical methods were proposed, such as observations of the moon. However, the computations would have been difficult for the typical sailor and the mathematical methods were not yet well-developed enough to provide an accurate prediction. This problem was among the foci of scientific activities of Galileo Galilei of Italy (1564–1642), who discovered the key property of pendulums that makes them useful for timekeeping: isochronism, which means that the period of swing of a pendulum is approximately the same for different sized swings. Galileo developed the idea for a pendulum clock in 1637, but did not have enough time to complete the design.

Dutch scholar Christian Huygens (1629–1695) successfully built a pendulum clock in 1656 and patented it the following year. Its design incorporated concepts derived from mathematical work on cycloids. The introduction of the pendulum—the first harmonic oscillator for timekeeping—increased the accuracy of clocks enormously, from about 15 minutes per day to 15 seconds per day. In addition to building a clock, Huygens investigated the properties of synchronization of identical pendulum clocks. Researchers have been interested in the subject of synchronization of clocks and oscillators since that time.

The design of the first marine chronometer was performed by the self-educated English carpenter and clockmaker John Harrison (1693–1776). This device dramatically revolutionized and extended the possibility of safe long-distance sea travel. At the time, the problem was considered so intractable that the British Parliament offered a prize of 20,000 British pounds sterling (comparable to about $4.72 million in modern currency) for the solution. Sailors and astronomers continued to be the principal consumers of accurate timekeeping. Precise clocks became essential equipment for each and every astronomical observatory.

Modern Clocks

The problem of "transportation" of accurate time to determine longitudes lost its actuality with the invention of the telegraph and, later on, with utilization of radio signals. But with the advancement of the twentieth century, new scientific and applied challenges demanded increasingly accurate time reckoning. As a result, new clocks were created based on newly discovered physical principles that were operationalized using mathematics. The crucial step in this direction was the invention of so-called quartz clocks. A quartz crystal has the unusual property of piezoelectricity—when stimulated with voltage and pressure, it oscillates at a constant frequency.

The vibration of a quartz crystal regulates the clock very precisely. Quartz crystal clocks were designed in 1927 by two engineers at Bell Telephone Laboratories: the Canadian-born telecommunications engineer Warren Marrison (1896–1980) and an electrical engineer from the Massachusetts Institute of Technology (MIT), Joseph Warren Horton (1889–1967). Since the 1970s, quartz clocks have become the most widely used timekeeping technology. Atomic clocks followed quartz clocks toward the end of the century. The U.S. National Bureau of Standards (now the National Institute of Standards and Technology) based the time standard of the land on quartz clocks between the 1930s and the 1960s. Eventually, it changed to atomic clocks, the best of which are accurate to 5×10^{-15} seconds per day. Researchers are now developing optical clocks that can be up to 100 times more accurate than the best atomic clocks. Further, satellite-based global positioning systems are now a primary source of time for some scientists and people in everyday life. This system provides almost unlimited transportation of time using variety of mobile devices in space and on Earth.

Today, the reckoning and keeping of precise and super-precise time continues to be requisite for numerous scientific and applied problems. Astronomers are still important users of this data. It is important, for instance, in cosmic navigation, in the measurement of variations of the rotation of Earth, and in the implementation of a particular technology into everyday life, such as radio interferometry with a hyperlong base. Every developed country now has a specialized national service for addressing questions regarding precise timekeeping and time reckoning. For a long time in the

Paris Observatory, there was the Bureau International de l'Heure (The International Time Bureau), which played an important role in the research of timekeeping. In 1987, the responsibilities of the Bureau were taken over by the International Bureau of Weights and Measures (BIPM) and the International Earth Rotation and Reference Systems Service (IERS).

Further Reading

Bruton, Eric. *The History of Clocks and Watches.* New York: Time Warner Books, 2003.

Collier, J. L. *Clocks.* New York: Benchmark Books, 2004.

Landes, David S. *Revolution in Time: Clocks and the Making of the Modern World.* Cambridge, MA: Belknap Press of Harvard University Press, 2000.

Sobel, Dava. *Longitude: The True Story of a Lone Genius Who Solved the Greatest Scientific Problem of His Time.* New York: Walker, 1995.

Uresova, Libuse. *European Clocks. An Illustrated History of Clocks and Watches.* London: Peerage Books, 1986.

Alexander A. Gurshtein

Coding and Encryption

Category: Communication and Computers.
Fields of Study: Algebra; Data Analysis and Probability; Number and Operations; Representations.
Summary: Mathematical algorithms are used in modern encryption and decryption.

Human beings have a propensity to preserve and share secret information. Cryptography, from the Greek *kryptos* (hidden) and *graphein* (to write), is the art and science of coding and decoding messages containing secret information. Encryption is the algorithmic process that converts plain-text into cipher-text (looks like a collection of unintelligible symbols), while decryption is the reverse process that converts the cipher-text back to the original plain-text. A cipher algorithm and its associated key control both directions of the sequence, with the code's security level directly related to the algorithm's complexity. The two fundamental types of cryptography are symmetric (or secret keys) or asymmetric (or public-key), with multiple variations. Claude Shannon, an American mathematician and electronic engineer, is known as the father of information theory and cryptography. Some claim that his master's thesis, which demonstrates that electrical applications of Boolean algebra can construct and resolve any logical numerical relationship, is the most important master's thesis of all time.

Around 2000 B.C.E., Egyptian scribes included nonstandard hieroglyphs in carved inscriptions. During war campaigns, Julius Caesar sent coded information to Roman generals. Paul Revere's signal from a Boston bell tower in 1775 is even a simple example of a coded message. Success of the Allies in both World Wars depended on their breaking of the German's Enigma code. With the world-wide need for more sophisticated coding algorithms to transmit secure messages for military forces, businesses, and governments, people began capitalizing on the combined powers of mathematics, computer technology, and engineering.

The simplest examples of ciphers involve either transpositions or substitutions. In 450 B.C.E., the Spartans used transposition ciphers when they wound a narrow belt spirally around a thick staff and wrote a plain-text (or message) along the length of the rod. Once unwound, the belt appeared to be a meaningless sequence of symbols. To decipher the cipher-text, the receiver wound the belt around a similar staff. Variations of transposition ciphers are the route cipher and the Cardan grill.

Julius Caesar used substitution ciphers, where each letter of the plain-text is replaced by some other letter or symbol, using a substitution dictionary. For example, suppose:

Original Alphabet:
A B C D E F G H I J K L M N O P Q R S T U V W X Y Z
Key Dictionary:
K L M N O P Q R S T U V W X Y Z A B C D E F G H I J

where the key dictionary is made by starting with "code" letter K and then writing the alphabet as if on a loop. To encode the plain-text, "The World Is Round," each letter is substituted by its companion letter, producing the cipher-text "CRO FXAUN SB AXDWN." To disguise word lengths and to add complexity, the cipher-text was sometimes blocked into fixed-length groups of letters such as "CROF XAUN SBAX DWN." To decipher the cipher-text, one needed to know only the "code" letter. Though simple and initially confusing, substitution ciphers now are easily broken using frequency patterns of letters and words. Variations of the substitution cipher involve the suppression of letter frequencies, syllabic substitutions, or polyalphabetic substitutions such as the Vigenère or Beaufort ciphers.

The Playfair Square cipher used by Great Britain in World War I is a substitution cipher, but its encryption of letter pairs in place of single letters is more powerful yet easy to use. The cipher-key is a 5 × 5 table initiated by a key word, such as "mathematics."

M	A	T	H	E
I	C	S	B	D
F	G	K	L	N
O	P	Q	R	U
V	W	X	Y	Z

The table is built by moving left to right and from top to bottom (or other visual pattern as in a spiral) by first filling in the table's cells with the keyword's letters—avoiding duplicate letters. Then, the subsequent cells are filled with the remaining letters of the alphabet, using the "I" to represent the "J" to reduce the alphabet to 25 letters (instead of 26). Both the coder and the decoder need to know the both the keyword and the conventions used to construct the common cipher-key.

The coder first breaks the plain-text into two-letter pairs and uses the cipher-key via a system of rules:

- If double letters occur in the plain-text, insert an X between them.
- Rewrite the plain-text as a sequence of two-letter pairs, using an X as a final filler for last letter-pair.
- If the two letters lie in the same row, replace each letter by the letter to its right (for example, CS becomes SB).
- If the two letters lie in the same column, replace each letter by the letter below it (TS becomes SK and PW becomes WA).
- If the two letters lie at corners of a rectangle embedded in the square, replace them by their counterpart in the same rectangle (TB becomes SH and CR becomes PB).

Using this cipher-key, the plain-text "The World Is Round" becomes first

TH EW OR LD IS RO UN DX

which when encoded, becomes

HE ZA PU BN CB UP ZU ZS.

The same cipher-key is used to decode this message, but the rules are interpreted in reverse. It is quite difficult to decode this cipher-text without access to both the keyword and the conventions to construct the common cipher-key, though very possible.

The problem with all substitution and transposition encryption systems is their dependence on shared secrecy between the coders and the intended decoders. To transmit plain-text via cipher-text and then decode it back to public-text successfully, both parties would have to know and use common systems, common keywords, and common visual arrangements. In turn, privacy is required, since these systems are of no value if the user learns the key-word or is able to use frequency techniques of word/letter patterns to break the code. A more complicated and secure encryption process was needed, but it was not invented until the 1970s.

The revolutionary idea in encryption was the idea of a public key system, where the encryption key is known by everyone (that is, the public). However, the twist was that this knowledge was not useful in figuring out the decryption key, which was not made public. The RSA public-key cipher, invented in 1977 by Ronald Rivest, Adi Shamir, and Leonard Adleman ("RSA" stands for the names of the inventors), all of whom have bachelor's degrees in mathematics and advanced degrees in computer science, is still used today thanks to powerful mathematics and powerful computer systems.

In a RSA system, the "receiver" of the intended message is the driver of the process. In lieu of the "sender," the receiver chooses both the encryption key and the matching decryption key. In fact, the "receiver" can make the encryption key public in a directory so any "sender" can use it to send secure messages, which only the "receiver" knows how to decrypt. Again, the latter decryption process is not even known by the "sender."

Because the problem is quite complex and uses both congruence relationships and modular arithmetic, only a sense of the process can be described as follows:

- As the "receiver," start with the product n equal to two very large prime numbers p and q.
- Choose a number e relatively prime to $(p-1)(q-1)$.
- The published encryption key is the pair (n,e).
- Change plain-text letters to equivalent number forms using a conversion such as $A = 2, B = 3, C = 4, \ldots, Z = 27$.
- Using the published encryption key, the "sender" encrypts each number z using the formula $m \equiv z^e \mod(n)$, with the new number sequence being the cipher-text.
- To decode the text, the "receiver" not only knows both e and the factors of n but also the large primes p and q as prime factors of n.
- Then, the decryption key d is private but can be computed by the "receiver" using an inverse relationship $ed \equiv 1 \mod (p-1)(q-1)$, which allows the decoding of the encrypted number into a set of numbers that can be converted back into the plain-text.

The RSA public key system works well, but the required primes p and q have to be very large and often involve more than 300 digits. If they are not large, powerful computers can determine the decryption key d from the given encryption key (n,e) by factoring the number n. This decryption is possible because of the fact that, while computers can easily multiply large numbers, it is much more difficult to factor large numbers on a computer.

Regardless of its type, a cryptographic system must meet multiple characteristics. First, it must reflect the user's abilities and physical context, avoiding extreme complexity and extraneous physical apparatus. Second, it must include some form of error checking, so that small errors in composition or transmission do not render the message into meaningless gibberish. Third, it must ensure that the decoder of the cipher-text will produce a single, meaningful plain-text. There are many mathematicians working for government agencies like the National Security Agency (NSA), as well as for private companies that are developing improved security for storage and transmission of digital information using these principles. In fact, the NSA is the largest employer of mathematicians in the United States.

Further Reading

Churchhouse, Robert. *Codes and Ciphers*. Cambridge, England: Cambridge University Press, 2002.

Kahn, David. *The Codebreakers: The Story of Secret Writing*. New York: Macmillan, 1967.

Lewand, Robert. *Cryptological Mathematics*. Washington, DC: Mathematical Association of America, 2000.

Smith, Laurence. *Cryptography: The Science of Secret Writing*. New York: Dover Publications, 1971.

Jerry Johnson

Communication in Society

Category: School and Society.
Fields of Study: Communication; Connections.
Summary: Communication helps mathematicians and others be informed of past and current research and to formulate and organize their own ideas.

Communication is fundamental to mathematics as a discipline, the mathematics community, mathematics education, and society as a whole, since communication is an essential part of everyday life and any social interaction. Effective communication is inherent in validating mathematics. Using a common language and a set of notions and drawing upon a shared body of knowledge, mathematicians communicate with each other—both orally and in writing—about their mathematical ideas, perceptions, or methods.

For example, mathematicians exchange ideas with their colleagues, write technical reports, publish original research papers and expository articles in professional journals, or give oral presentations. Some associate good mathematics communication with beautiful expository lectures or clear writing, while others focus on the quality of the interactions between people, such as those working in a group on mathematics. A peer review process is frequently part of mathematics communication and dissemination, ensuring some degree of consensus on what constitutes *appropriate* or *valid* mathematics. In this way, the standards of mathematics are socially developed. In addition to interacting with their colleagues, mathematicians need to communicate with the rest of the society using a language and terminology that are more familiar to the general public. For instance, mathematicians explain to the public how the discipline of mathematics contributes to society or demonstrate the various applications of mathematics in fields such as engineering, medicine, and communication technologies.

The role of communication in the education of mathematics is similar to the vital role communication plays in the discipline of mathematics. Drawing upon mathematical language and notation, teachers and students talk about mathematics; share, explain, and justify mathematical ideas; or analyze, discuss, and interpret mathematical concepts. Communication about mathematics and communication using mathematical language do not occur only in the mathematics community or in mathematics classrooms. Regardless of one's profession, wise decision making in personal lives and participation in civic and democratic life increasingly demand mathematical communication skills. For example, people need to communicate with mortgage companies when buying a house and interpret various mathematical concepts (such as percentage and rate) presented in the media. Thus, communication with mathematics and about mathematics is an essential part of daily life.

Communication Media

In the twenty-first century, there are a wide variety of electronic and print venues for communicating mathematics, and the evolution of electronic media and databases has vastly changed the way people access mathematics. Historically, mathematicians communicated by letters, during visits, or by reading each other's published articles or books once such means became available. Some mathematical concepts were developed in parallel by mathematicians working in different areas of the world, such as German Karl Friedrich Gauss and American Robert Adrain, who both made advances in the theory of the Normal distribution in the early nineteenth century. Some mathematicians were not aware of each other's progress because they did not have the venues of communication that are available in the twenty-first century. In an effort to increase the accessibility of mathematics research articles, reviews began appearing in print journals like *Zentralblatt für Mathematik*, which originated in 1931, and *Mathematical Reviews*, which originated in 1940. Since the 1980s, electronic versions of these reviews have allowed researchers to search for publications on a specific topic. In 2010, MathSciNet, the electronic version of *Mathematical Reviews*, listed more than 2 million items and more than 1 million links to original articles. In 2011, the database Zentralblatt MATH listed more than 3 million items from approximately 3500 journals and 1100 serials. Both contain work dating back to the early 1800s. There are also thousands of mathematics journals that are not listed in these collective databases, such as most mathematics education research.

Some mathematicians publish open access drafts of their papers on their personal Web pages before official publication in peer-reviewed and other journals, or in other online settings such as the ArXiv.org e-print archive. Co-authors from around the world can work together using e-mail or other Web-based collaborative tools. Mathematics students, teachers, and researchers

National Council of Teachers of Mathematics

The National Council of Teachers of Mathematics (NCTM) emphasizes clear and coherent communication of mathematical ideas and thinking as a skill that students need to learn from pre-kindergarten through grade 12. Given the essential role of communication in teaching and learning of mathematics, NCTM has set forth process standards for communication for primary and secondary mathematics curricula. The Principles and Standards for School Mathematics (2000) states that instructional programs from pre-kindergarten through grade 12 should enable all students to

- Organize and consolidate mathematical thinking through communication
- Communicate their mathematical thinking coherently and clearly to peers, teachers, and others
- Analyze and evaluate the mathematical thinking and strategies of others
- Use the language of mathematics to express mathematical ideas precisely

often discuss mathematics ideas and share resources on blogs, through online chats, or using other forums. For instance, what began in 1992 as the Geometry Forum was extended in 1996 to become the Math Forum. There are many additional resources for sharing and teaching mathematics content, both in print and in electronic media. Some electronic examples include the National Council of Teachers of Mathematics Illuminations Web site; Wolfram MathWorld, which was developed by Eric Weisstein; and Math Fun Facts, developed by Francis Su. Social and historical context is also often addressed in sites such as The MacTutor History of Mathematics archive, developed by John O'Connor and Edmund Robertson, or Mathematicians of the African Diaspora, created by Scott Williams.

One important question related to online communication is how to represent and display mathematical notation, which is an important part of mathematical validity and understanding. Some Web pages contain fixed images for each equation or graph. Others use Java applets for dynamic display. The Mathematical Markup Language (MathML) is one way to encode mathematics. TeX was created by Donald Knuth in order to typeset scientific and mathematical research. TeX-based software such as LaTeX has become the standard in printing mathematics. Another issue is the validation of online resources, which may be created or published without peer review. On one level, this issue is an extension of the existing issue of peer review for print media, as mathematics journals already employ varying degrees of rigor when reviewing and publishing papers. At the same time, there is in increasing trend of creating printed works from electronic sources or using electronic sources as references, which creates an added difficulty in ensuring the collective accuracy of the body of mathematics communication.

With so many options available, the specific nature of mathematics communication depends in large part on the purpose and intended audience. There are some mathematics publications and communications aimed at a general audience, others aimed at students, and yet others intended for researchers. Mathematicians, educators, and other communications specialists work to match the form and venue of the mathematics communication to the need. Some careers that are regularly involved in communicating mathematics include technical writers or publication editors. The Society for Technical Communication and the Council of Science Editors are two professional associations that address this need. In 2007, Ivars Peterson became the director of Publications and Communications at the Mathematical Association of America, which, like other professional associations, publishes items for both the specialist and the nonspecialist. He previously wrote MathTrek for *Science News*. In 1991, he received a Joint Policy Board for Mathematics (JPBM) Communications Award for his "exceptional ability and sustained effort in communicating mathematics to a general audience." He also served as East Tennessee State University's Basler Chair of Excellence for the Integration of the Arts, Rhetoric, and Science in 2008 and taught a course there called Communicating Mathematics. In a talk on the topic of communication in mathematics, he noted:

The importance of communicating mathematics clearly and effectively is evident in the many ways in which mathematicians must write, whether to produce technical reports, expository articles, book reviews, essays, referee's reports, grant proposals, research papers, evaluations, or slides for oral presentations.

Communication in Schools

Communication, both oral and written, is an essential part of mathematics education. The act of communication allows students to systematize and incorporate their mathematics thinking and understanding, both for learning mathematical theory and mathematical problem solving. For example, when students communicate their own mathematical thinking and understanding, they are required to rationalize and organize their reasoning and also formulate puzzling or complex questions well enough to present them as clearly as possible to a reader. As a result, the process guides students toward greater insights to their own thinking and learning. Focused reflection, which is conceptually intertwined with communication, helps students to increase the benefits of communicating their ideas with peers, teachers, and others. Written or oral reflections in which ideas are shared among peers, teachers, and others provide students multiple perspectives that sharpen ideas explored. The American Society of Mathematics (ASM), which is also known as the American Society for the Communication of Mathematics, sponsors problem-solving contests and the U.S. National Collegiate Mathematics Championship.

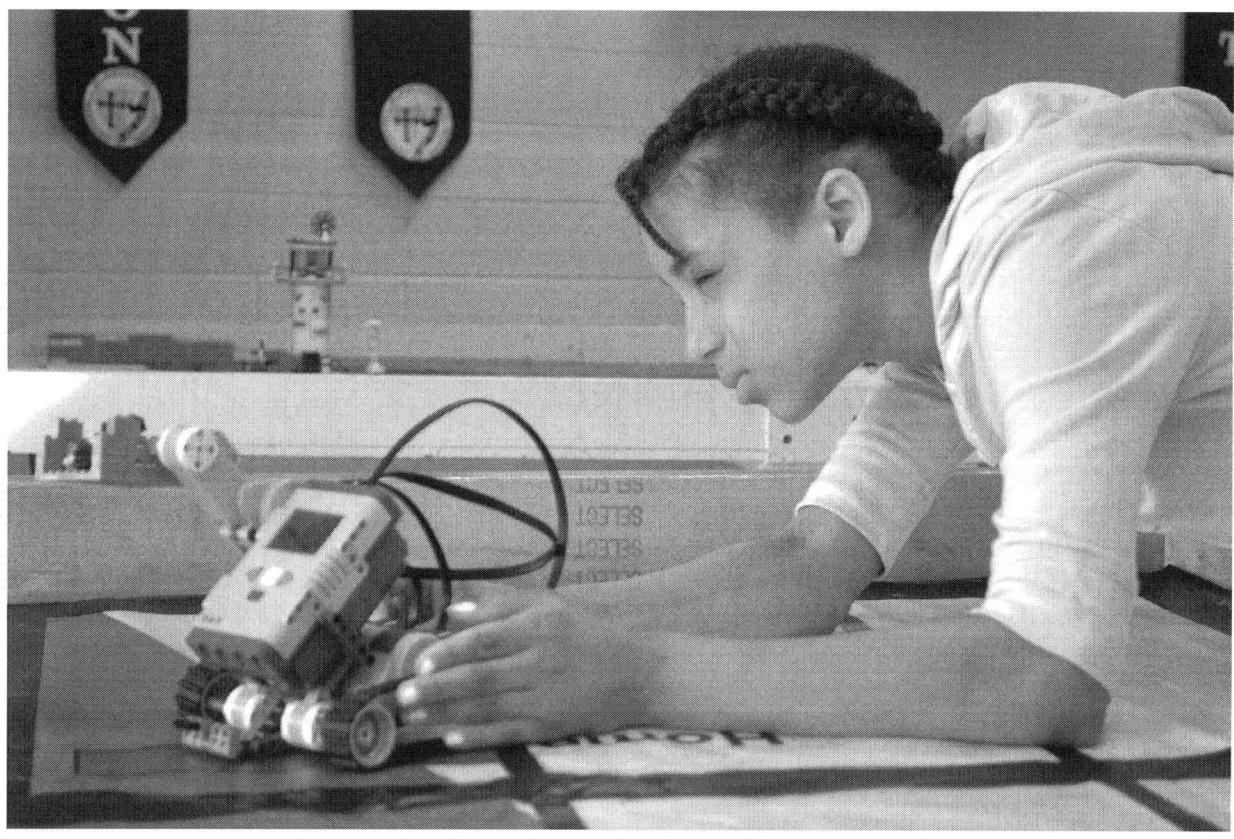

A student at a science, technology, engineering, and math (STEM) summer camp examines her robot before releasing it for a test. The camp encouraged K–12 students to pursue education and careers in the STEM fields. (U.S. Navy)

Proofs

One topic that illustrates the importance and the diverse nature of mathematics communication is the notion of proof. Researchers have proposed a wide variety of roles for proof in mathematics, such as establishing the truth of a statement, communicating mathematical knowledge, opening the way for further understandings and discoveries in mathematics, providing new techniques for doing mathematics, and organizing statements into systems of axioms and theorems. Throughout history, proofs and communication via proof have been incorporated in many different ways in mathematics education in the United States. The National Council of Teachers of Mathematics' (NCTM) 2000 Principles and Standards for School Mathematics emphasized the role of proof in mathematics learning for all students and helped to formalize its curricular importance and place in pre-kindergarten though high school education. Further, as proof became more systematized in K–12 education, some mathematics education researchers began to more deeply explore students' understanding of the definition or nature of proof, the role of proof as a mode of communication, and peer acceptance of the validity of a given proof, as well as how proof is taught in classrooms.

As the concept of proof came under investigation, an important issue was the conceptualization and the roles of proof in school mathematics. The NCTM defined proof in *Principles and Standards for School Mathematics* as "arguments consisting of logically rigorous deductions of conclusions from hypotheses." One element in the definition of proof is the acceptability of an argument as proof, which is referred to as "logically rigorous." An important question that NCTM's definition entails is who decides if a proof is logically rigorous enough to be accepted. To conceptualize the definition and identify the roles of proof in school mathematics, mathematics education researchers have referred to the qualifications and function of proof in the discipline of mathematics and investigated how it is implemented in mathematics classrooms.

Research has demonstrated the social nature of argumentation and justification in the classroom and beyond, and communication and validation by peers plays an important role in proof within and outside the classroom. This social dimension of proof is grounded in sociocultural theories of mathematical learning and is believed to reflect the process of becoming a mathematician. Yu Manin argued that within mathematics community, "A proof becomes a proof after the social act of 'accepting it as a proof.'" Erna Yackel and Paul Cobb concluded that acceptable justifications in mathematics education are interactively constituted by individual teachers and students in each classroom, where the teacher is the representative of the mathematical community. Mathematical justifications and argumentations are regulated by the general expectations and the regulations of the classroom community.

Thus, they are a part of the classroom norms and, more specifically, the sociomathematical norms, which are the extension of general classroom social norms to specifically focus on the normative aspects of mathematical discussions as students participate in mathematical activities. Yackel and Cobb argued that: "Normative understandings of what counts as mathematically different, mathematically sophisticated, mathematically efficient, and mathematically elegant in a classroom are sociomathematical norms.…The understanding that students are expected to explain their solutions and their ways of thinking is a social norm, whereas the understanding of what counts as an acceptable mathematical explanation is a sociomathematical norm." This idea plays a role in mathematics educator Andreas Stylianides's conceptualization of proof. He proposed four aspects that are required to consider an argument a proof: foundation, formulation, representation, and social dimension. He presented an example in which an elementary school student constructed a mathematical argument that was founded on definitions of mathematical constructs, formulated using deductive reasoning from these definitions, and then represented verbally. Regarding the social dimension of the proof, although the student's argumentation was logically rigorous and would have been accepted as a proof in the wider mathematical community, it generated counterarguments among her classroom peers and her argument was not accepted as a proof by the classroom community.

Indeed, the conceptualization of mathematics, in particular the social dimension that is appropriate for school mathematics, requires more research to develop. Mathematical discourse is an important factor in the development of shared understanding of mathematically valid justifications. However, students at various levels, particularly younger elementary school students, may have different levels of understand regarding the rules and norms of mathematical discourse, and understanding is not necessarily shared by all. Thus, as was the case in Stylianides' study, a valid mathematical argument was not accepted as valid by all students. In such cases, the teacher, acting as an authoritative representative of the mathematical community, could intervene and explain why the argument is indeed valid by broader standards. However, in some ways this action would negate the social dimension aspect that is used to evaluate mathematical acceptability, at least with respect to the classroom environment. Thus, the subtleties in what constitutes a valid argument within a mathematics classrooms and the relation to a teacher's role as the communicator of other mathematical norms as they acculturate students in the processes of proving need to be explored. It is important to note that teachers need to know when, how, and how much to intervene so as to not play an authoritarian role, thereby creating a learning environment in which students are forced into authoritarian schemes and communication is essentially unidirectional, from teacher to student.

Mathematical Applications in Communication Technologies

In the increasingly digital world of the twenty-first century, the safe communication of information has become a major issue for discussion and research in mathematics and science, in large part because of theft and fraud often perpetrated using new technologies. Mathematics plays an important role in making communication as safe as possible. Cryptology is a technique used to ensure that messages or data are transmitted safely to the receiver. Dating to the substitution ciphers used in ancient Rome and other civilizations, this field has always drawn heavily from mathematics. Research in mathematics and other disciplines, such as computer sciences and engineering, has resulted in an increasingly sophisticated array of coding techniques and technologies, as well as code-breaking methods. Some of the most common and known applications of cryptography include encryption of credit card numbers or passwords for electronic commerce and encryption of e-mail messages for secure communication. Confidentiality, authenticity, and integrity in electronic commerce or communication have become an apparent and sensitive issue for people who engage in online transactions such as buying or selling items online, online banking, and online communications, as well as for applications like medical records. If proper action is not taken for data transmission, information sent over an open network can be stolen by hackers. Such an action can reveal secret information or messages containing personal information, like a credit card number, a password, or online banking information, facilitating crimes like identity theft. A hacker can use digital data to clone a person's identity and use a victim's resources for the hacker's own good. Even worse, this information could be a national secret, and it may cause more serious problems. For that reason, the National Security Agency (NSA) uses its cryptologic heritage in the midst of challenging times to protect national security systems, and the NSA is one of the leading employers of mathematicians in the United States at the start of the twenty-first century.

Along with digital security, mathematics also plays a fundamental role in both the hardware and software that make the increasingly wireless, globally connected world possible. The *Advances in Mathematics of Communications* journal publishes research articles related to mathematics in communication technologies. Mathematicians and mathematical methods contribute to many aspects, including the Internet's computer server backbone and communications protocols; vast cell phone networks; and smartphones that act as mobile platforms for an array of communications methods, such as voice, text, photo, e-mail, and Internet. Music, movies, dance, art, theater, and many other methods people use to convey ideas to one another involve mathematics as part of the creative endeavor. Humans can communicate with neighbors next door, with people on the opposite side of the world, with satellites orbiting the planet, or even with probes that have been sent into the far reaches of the solar system thanks to mathematics. Some would in fact argue that mathematics is itself a universal language or method of communication.

Further Reading

Elliott, Portia, and Cynthia Garnett. *Getting Into the Mathematics Conversation: Valuing Communication in Mathematics Classrooms—Readings From NCTM's School-Based Journals*. Reston, VA: National Council of Teachers of Mathematics, 2008.

Manin, Yu. *A Course in Mathematical Logic*. New York: Springer-Verlag, 1977.

Mathematical Association of America. "JPBM Communications Award." http://www.maa.org/Awards/jpbm.html.

National Council of Teachers of Mathematics. *Principles and Standards for School Mathematics*. Reston, VA: National Council of Teachers of Mathematics, 2000.

Peterson, Ivars. "Writing Mathematics Well." http://sites.google.com/site/ivarspeterson/workshop1.

Stylianides, Andreas. "The Notion of Proof in the Context of Elementary School Mathematics." *Educational Studies in Mathematics* 65 (2007).

Yackel, E., and P. Cobb. "Sociomathematical Norms, Argumentation, and Autonomy in Mathematics." *Journal for Research in Mathematics Education* 22 (1996).

Zeynep Ebrar Yetkiner Ozel
Serkan Ozel

Dams

Category: Architecture and Engineering.
Fields of Study: Algebra; Geometry; Measurement.
Summary: Mathematics is vital to the design, monitoring, maintenance, and safety of dams.

Dams are embankments across a waterway for control of water or for water storage; they have served many functions in societies throughout history. The earliest dams were primarily used for irrigation and as a water source for livestock. Today, smaller dams provide water for livestock, fish and wildlife habitat, and recreation. Larger dams can provide flood control in places below sea level, like New Orleans and the Netherlands; municipal and industrial water supply; irrigation for crops; hydroelectric power; commercial navigation; and recreation. They are typically earthen dams, concrete structures, or some combination. Older dams were sometimes made of timber, masonry, or steel. Mathematicians and engineers investigate many aspects of the construction and maintenance of dams using geometry, trigonometry, and stochastic and limit-state analyses. For instance, Boris Galerkin, who had degrees in applied mathematics and mechanics, studied stress in dams, and Pelageia Polubarinova, who had a degree in mathematics, contributed to the theory of seepage flow of groundwater through porous materials that included earth dams. Some well-known dams are the Itaipú Dam in Brazil and Paraguay, the Hoover Dam in the United States, the Aswan Dam in Egypt, and the Dneproges Dam in the Ukraine.

Considerations for building a dam must take into account both positive and negative impacts. There are a variety of benefits of a dam that are closely related to its uses—providing water supply, flood control, hydroelectric power, and navigation. Hydroelectric power provides an important source of electrical power around the world. Commercial navigation through river systems provides efficient and economical transportation of agricultural products and commercial goods. Many dams that control flood plains provide farmers with an increased crop yield because land that would once have been flooded is now controlled upstream by the dam. Negatively, some dams may hinder fish movement; for example, along some streams, salmon are not able to get back to their native spawning areas because of the dam. Additionally, dams affect the natural order of a stream—its sediment load and flooding characteristics.

Purposes and Design

Dams are constructed with a definite purpose in mind based on the function(s) they are to serve. Dams are built to control watershed areas (all the area upstream of the dam, which provides runoff to the structure). Engineers use a variety of mathematics skills as they plan, design, construct, and operate a dam. During the planning stage, engineers work with sponsors to scope out the needs and develop a basic design for the structure including design issues such as location, height, and base flow of the structure. Base flow is calculated with the formula $Q = v \times A$ where Q is the base flow rate, v is the velocity of water, and A is the area. Another important part of the planning stage is determining the economic feasibility of building the dam by calculating a benefit-to-cost ratio. Using a mathematical model, both the benefits of the dam over its life and the total cost of building and maintaining the dam are calculated. Ideally, for the construction of a dam to be feasible, the benefit-to-cost ratio needs to be greater than 1.

As a part of the design process, engineers must create detailed blueprints for the structure and an accompanying cost sheet that includes items such as quantities or volumes of a variety of materials (for example, cubic yards of concrete) and the cost of the removal and placement of earthen materials, which can be millions of cubic yards in the case of large dams. During the construction of the dam, the blueprints must be followed with precision and detail to ensure the integrity of the dam. Once the dam is constructed, regular monitoring is important to ensure the most efficient use of the available storage. Engineers monitor the amount of water leaving the dam through its spillway, as well as the amount of water entering the watershed. These inflows and outflows must be balanced in order to maintain storage needs and prevent flooding or low flows in the river downstream.

Safety

A major consideration in the planning, design, construction, and maintenance of any dam is safety. Engineers determine a hazard rating for each dam, with the highest hazard rating dealing with potential loss of human life. A breach in a dam can be catastrophic. A breach in a dam can be caused by a flaw in the design of the structure, extreme rainfall, lack of or poor maintenance of the structure, or a geological occurrence. Regular inspection and maintenance are important to ensure the safety of those downstream from the dam.

Further Reading

Hiltzik, Michael. *Colossus: Hoover Dam and the Making of the American Century*. New York: Free Press, 2010.

Macy, Christine. *Dams*. New York: W. W. Norton & Company, 2009.

Prabhu, N. U. *Stochastic Storage Processes: Queues, Insurance Risk, Dams, and Data Communication*. 2nd ed. New York: Springer, 1998.

Juliana Utley

Deep Submergence Vehicles

Category: Travel and Transportation.
Field of Study: Algebra; Measurement; Number and Operations.
Summary: Submergence vehicles must be carefully designed to take into account undersea conditions.

Deep submergence vehicles are primarily designed to aid researchers in exploring the depths of Earth's oceans. Much is unknown about the suboceanic environment, and exploration of these depths requires transport vehicles that can withstand tremendous pressures. Modern submergence vehicles can not only dive to great depths but can also stay submerged for hours at length, and are equipped with external lights and tele-operated robotic manipulators to gather deep sea samples for further research. Besides researching marine life, deep submergence vehicles also play vital roles in the oil exploration and the telecommunications industries where robotic submarine vehicles known as "autonomous underwater vehicles" detect faulty cables and help in oil field exploration. English mathematician William Bourne may have been the first to record a design for an underwater navigation vehicle in 1578. In addition to mathematics and mathematicians impacting deep submergence vehicles, submarines have also impacted the development of mathematics. Mathematicians examined the optimal way for airplanes to search for submarines, and the field of operations research was born.

Physical Characteristics of the Abyss

Pressure. At any given depth under the sea level, the pressure on a body can be calculated as

$$P = \rho \times g \times h$$

where P is pressure, ρ is the density of the seawater, g is the acceleration because of gravity, and h is the depth at which the measurement is being taken.

The atmospheric pressure at sea-level is about 100 kPa (~ 14.6 psi), the same amount of water pressure at about 10 meters (33 feet) below the surface, making the combined pressure experienced by a body at a 10 meter depth almost double of that at the surface.

Light. Most of the visible light entering the ocean is absorbed within 10 meters (33 feet) of the water's surface. Almost no light penetrates below 150 meters (490 feet). Solid particles, waves, and debris in the water affect light penetration. The longer wavelengths of light, red, yellow, and orange, penetrate to 15, 30, and 50 meters respectively, while the shorter wavelengths—violet, blue, and green—can penetrate further. The depth of water where sunlight penetrates sufficiently for photosynthesis to take place is called the Euphotic Zone and is normally around 200 meters (655 feet) in the ocean. The zone where filtered sunlight only suffuses in the water is known as the Disphotic Zone and extends from the end of the Euphotic Zone to about a depth of 1000 meters. Below that, no sunlight ever penetrates, and this is known as the Aphotic Zone.

Temperature. There is a significant difference in the temperatures between the Euphotic and Aphotic zones. However, in the Aphotic Zone, the temperature remains almost constant, hovering around 2 to 4 degrees Celsius. The only exception occurs when deep-sea volcanoes or hydrothermal vents exist, which cause significant warming of the waters.

History

The earliest deep-sea submersibles were known as "bathyspheres" (from *bathys*, Greek for "deep"). They were raised in and out of the water by a cable. They were fitted with oxygen cylinders inside to provide air to the divers, and had chemicals to absorb the expelled carbon dioxide. The early bathyspheres were not maneuverable—the only degree of freedom they had enabled them to go up and down.

The notable Swiss physicist Auguste Piccard (1884–1962) was influential in making the next design iteration to the bathysphere, called the "bathyscaph." The vessel was not suspended from a ship but instead attached to a free-floating tank filled with petroleum liquid. This tank made it buoyant (lighter than water). The bathyscaph had metal ballasts that, when released, allowed the vessel to surface. Auguste and his son Jacques designed the next generation bathyscaph, the Trieste. The Trieste set a new world record when it reached the lowest point on Earth, the Marianas Trench (35,800 feet).

Improvements in electronics and materials engineering have led to the design of Alvin, a deep-sea vessel capable of accommodating up to three people and diving for up to nine hours. Alvin sports two robotic arms that can be customized depending on the mission it is undertaking. Alvin's most notable contribution was its role in exploring the RMS Titanic.

Further Reading

Arroyo, Sheri, and Rhea Stewart. *How Deep Sea Divers Use Math.* New York: Chelsea House, 2009.

Morse, Philip and George Kimball. *Methods of Operations Research.* Kormendi Press, 2008.

Mosher, D. C., Craig Shipp, Lorena Moscardelli, Jason Chaytor, Chris Baxtor, Homa Lee, and Roger Urgeles. *Submarine Mass Movements and Their Consequences.* New York: Springer, 2009.

ASHWIN MUDIGONDA

Digital Book Readers

Category: Communication and Computers.
Fields of Study: Geometry; Measurement; Number and Operations; Representations.
Summary: The twenty-first-century surge in e-books began with the advent of "electronic ink" and future innovations include sketchpad-like functionality.

People have been reading digital content on computer screens since the 1970s, but the technology used for most computer screens at the end of the twentieth century made them somewhat less useful for replacing paper books, magazines, and newspapers. In 1971, volunteers started digitizing and archiving books for Project Gutenberg, whose goal was to encourage the development of electronic books. Research on electronic paper began in the 1970s. Many open and proprietary digital document formats were devised for potential use in e-books, like Adobe's Portable Document Format (PDF), created by mathematician and engineer John Warnock. However, most early attempts at digital books were unsuccessful or aimed at niche technical audiences.

In the early twenty-first century, the E Ink company introduced electronic ink technology, which revolutionized digital books. The company was co-founded by several individuals, including physicist Joseph Jacobson and Russell Wilcox, who has a degree in applied mathematics. The resulting "electronic paper" has a high contrast ratio similar to standard paper, and for most users it closely matches the experience of reading on standard paper. One early application was flexible, changeable store signs. The 2004 Sony Librié, released in Japan, was the first e-reader to make the technology widely available, while the Amazon Kindle is credited with popularizing it in the United States. As of 2010, there were many variations on e-readers with the ability to display multiple e-book formats. Some of the most popular included the Sony Reader, the Amazon Kindle, and the Barnes & Noble Nook. Motorola's FONE F3 was the first portable phone to include this technology.

Electronic Ink

Electronic ink technology is based on microcapsules, which were already in use for applications like scratch-and-sniff stickers and time-release medications. Rotating microcapsule spheres for electronic ink are filled with a clear liquid containing a mix of small, electrically charged black and white particles. Some implementations contain on the order of 100,000 spheres per square inch. Electronic paper is a sheet of plastic coated with millions of microcapsules and equipped with an electronic device to draw the black and white particles into desired patterns of black and white dots.

When viewed from a distance, the patterns create words and pictures. The dots can also be mixtures of black and white, resulting in a range of grayscale tones. To change the image, computer programs in the reader send an electronic pulse to rearrange the pattern. Microcapsules are bistatic, which means they stay in place once they are arranged without drawing continuous electrical power. This factor contributes to long battery life. Electronic paper also has no backlighting like personal computer screens; it uses light reflection for viewing, just like ordinary paper. Scientists are investigating red, green, and blue filters to produce full-color electronic ink images. A version of the Barnes & Noble Nook released in 2010 uses a liquid crystal display (LCD) screen for color and touch-screen functionality. Some praise this, while others consider it to be a step backward in e-reader technology.

Early twenty-first-century digital book readers embody several other features that make them well-suited alternatives for leisure reading and textbooks in schools. One important aspect is their portability with high-capacity storage. Typical readers have the capability to store hundreds of books, so all required textbooks could be stored in single digital reader. Connectivity via wireless networking allows the downloading of a variety of books or teacher-created documents, including RSS feeds for blogs and Web content. RSS was developed by programmers like David Winer, who has degrees in mathematics and computer science. The reading experience is customizable; some e-readers have touch-screen navigation, adjustable font levels, the ability to take notes directly on screen or highlight text sections, built-in dictionaries, or search functions.

Readers that debuted in 2010 featured applications to allow users to write or draw, like a tablet PC, which would be important for many mathematical subjects, like geometry. Some mathematics educators have explored the use of electronic ink to support mathematics distance education. For example, electronic ink tools in a chat program allowed students and instructors to post and edit mathematical formulas, diagrams, and graphs while communicating in real time.

Further Reading

Howard, Nicole. *The Book: The Life Story of a Technology*. Westport, CT: Greenwood Press, 2005.

Kipphan, Helmut. *Handbook of Print Media: Technologies and Production Methods*. New York: Springer, 2001.

Serkan Ozel
Zeynep Ebrar Yetkiner

Digital Cameras

Category: Arts, Music, and Entertainment.
Fields of Study: Geometry; Measurement; Number and Operations; Representations.
Summary: Rapid advances in digital camera technology have led to their widespread use.

From the invention of modern photography in the 1800s to the rise of digital photography in the twenty-first century, the function of the camera has been the same: to record patterns of light. The word "photography," coined by Sir John Herschel in 1839, is from the Greek *phos* (light) and *gráphein* (to write). Simple pinhole cameras were described as early as the fourth and fifth centuries B.C.E. by Chinese philosopher Mo Ti and Greek mathematicians Aristotle and Euclid of Alexandria. Mathematician and physicist James Maxwell created the first color photograph in 1861. Not long after, American inventor and Kodak founder George Eastman developed inexpensive equipment and film that made photography practical for common use. Until recently, cameras recorded images on media coated in photosensitive compounds. Incoming light was registered as a chemical change that could be seen upon development in specialized photochemistry. Digital cameras use an electronic chip that is sensitive to light. The chip, either a charge coupled device (CCD) or complementary metal-oxide-semiconductor (CMOS), converts the light into an electrical signal, and a small computer in the camera then transforms that signal into the "ons" and "offs" (or "1"s and "0"s) of binary code for storage on a digital storage device. The digital information that represents an image can easily be copied onto a computer, manipulated, published electronically, and printed. Researchers also investigate mathematical questions like how many images one should shoot in order to be reasonably confident that no person in the photograph blinks. For groups under 20 people, the number of images is approximately equal to the number of people divided by one-half.

The Lens and the Shutter

Like most cameras, a digital camera begins the process of taking a photograph by letting light in through a lens (a curved piece of glass or plastic) that bends light through the principle of refraction and focuses the image. The light then passes through an opening called an "aperture" whose size can be adjusted to let more or less light pass. Apertures are described by an f-stop number, which is proportional to the focal length of the lens over the diameter of the entrance pupil. Since it is a ratio, larger f-stop numbers refer to smaller apertures. Each doubling or halving of the f-stop number translates to a change in amount of light let in by a factor of four. Thus, an f-stop of 11 lets in four times more light than an f-stop of 22. Finally, the light comes to a shutter that opens for a period of time when the shutter release is triggered, allowing light into the camera body. Usually, a shutter speed is a fraction of a second, though long shutter speeds can be used in low light, or for a variety of effects. With especially long shutter speeds, heat can build up in the CCD or CMOS, causing electrical interference that interferes with accurate, binary recording of the image, resulting in error or "noise," though camera manufacturers are developing a number of processes that have made this less of a problem over time.

The CCD or CMOS

In order to capture an image, the light that comes into the camera falls on the CCD or CMOS chip, which changes the image into electric current. A CCD is made up of tiny regions called "picture elements" or "pixels" that will correspond to the points in the photograph. A CMOS works similarly, though the specific underlying technology is a bit different. Some cameras have one CCD for all three primary colors of visible light, red, green, and blue; each pixel records only color of light from the scene. More advanced cameras use three different CCDs, one for each primary color, resulting in a more accurate image. Ultimately, the electric current from the CCD is encoded by a small computer in the camera into a stream of binary information in the form of "ons" and "offs" that ultimately will be stored on a flash memory card.

Sensitivity and ISO

In film cameras, different formulations of film are used for different light conditions, with more sensitive films employed in low light. In a digital camera, the signal from the CCD can be boosted to handle low light levels; however, doing so introduces noise in the signal. The setting for camera sensitivity is described as its "ISO number," an international standard for measuring the speed of color film. It uses both an arithmetic and logarithmic scale to combine two previous film standards. In the arithmetic scale, which is commonly the only one given, each doubling of the ISO representing a doubling of the sensitivity. Thus, a camera set to ISO 100 will be half as sensitive to light—and will require twice as long an exposure for a given scene to achieve the same result, given the same f-stop setting—as one with an ISO of 200.

Pixel Dimensions

One of the factors that determines the picture quality of an image produced by a digital camera is the number of pixels it records. This is especially relevant when images are blown up to large dimensions, as the individual pixels begin to become visible. Pixels are the individual binary units into which the image is "broken up" and stored during the electronic conversion process by the camera's chip. For example, the Droid Incredible phone, released in 2010, contains an 8 megapixel camera, which means its photographs are composed of about 8 million individual pixels, with each picture having a possible resolution of roughly 3264 pixels wide by 2468 pixels high. However, in practice, there are many factors that affect picture quality. The size of the electronic chip plays a large role. When the photosensitive regions of a camera's chip are packed too tightly together, they create electronic interference in their neighbors, potentially affecting the binary storage, and ultimately affecting the accuracy and quality of the stored image.

Further Reading

Stone, J., and B. Stone. *A Short Course in Digital Photography.* Upper Saddle River, NJ: Prentice Hall, 2009.

Cameras in Mathematics Classrooms

Cameras have grown in popularity since Eastman first made them readily available. Digital cameras are relatively inexpensive, and, in fact, are now standard features on many cell phones. Educators have seized on the digital camera as a very useful tool in the classroom for introducing concepts, making connections, and enriching educational experiences in a very hands-on way. For example, students in middle grades and above have been asked to use digital cameras to record their own examples of geometric concepts found in the world. They can then use the photographs, along with various mathematics concepts such as scaling and trigonometry, to answer questions like "Is the Houston Astrodome really round?" or "What is the slope of a roof?" In other cases, students use photos to record and measure themselves and their classmates—either once or repeatedly over time—to provide data for many interesting mathematics activities and discussions, such as variability and the importance of repeated sampling.

Svenson, Nic. "Velocity Science in Motion: Blink-free Photos, Guaranteed." http://velocity.ansto.gov.au/velocity/ans0011/article_06.asp.

Jeff Goodman

Digital Images

Category: Arts, Music, and Entertainment.
Fields of Study: Algebra; Geometry; Number and Operations; Representations.
Summary: Digital images are recorded as a binary account of pixels, which algorithms may compress.

Digital images are not images at all but rather are visual information encoded as binary data. Viewing a digital image requires a computer to decode binary information and display it on a screen in the form of an array of discrete lights called "picture elements" or "pixels." The first computer-generated digital images were produced in the early 1960s. The needs of the Cold War, medicine, and the space race drove many developments in digital imagery, some of which were achieved in the context of projects on satellite imagery, medical imaging, optical character recognition, and photo enhancement. The advent of microprocessors in the 1970s and advances in digital storage and display technologies made possible sophisticated imaging tools, like computerized axial tomography (CAT scanning).

The degree of mathematical sophistication that CAT scans introduced into medical imaging, such as integral geometry, optimal sampling, and transport equations, was unheard of at that time. It is reflected in further advances such as magnetic resonance imaging as well as developments in other fields that use similar imaging techniques, like seismic and electron microscopy. At the same time, scanners to digitize analog images began to be used in a diverse array of fields, such as archaeology and law enforcement. The first fully digital camera was released in 1995, and by the end of the twentieth century, charge-coupled devices (CCDs) largely displaced analog film and tape for photography and videography. Willard Boyle and George Smith shared the 2009 Nobel Prize in Physics for their invention of the CCD, an idea they first brainstormed at Bell Labs in 1969. Improved computing power also allowed for production of near-photorealistic images. All areas of digital imagery (creation, compression, restoration, recognition, and display) involve mathematics. In the twenty-first century, digital images are regularly used in both mathematics research and teaching.

Bitmap Graphics

In most digital images, each pixel has been defined numerically and this number has been converted into a string of "1"s or "0"s. This system is the approach of "bitmap graphics" (also known as "raster graphics"), and it is how digital cameras work. Depending on the number of bits used to represent each pixel, more or less color information is given. For example, a one-bit system would allow only a black or white pixel, as the

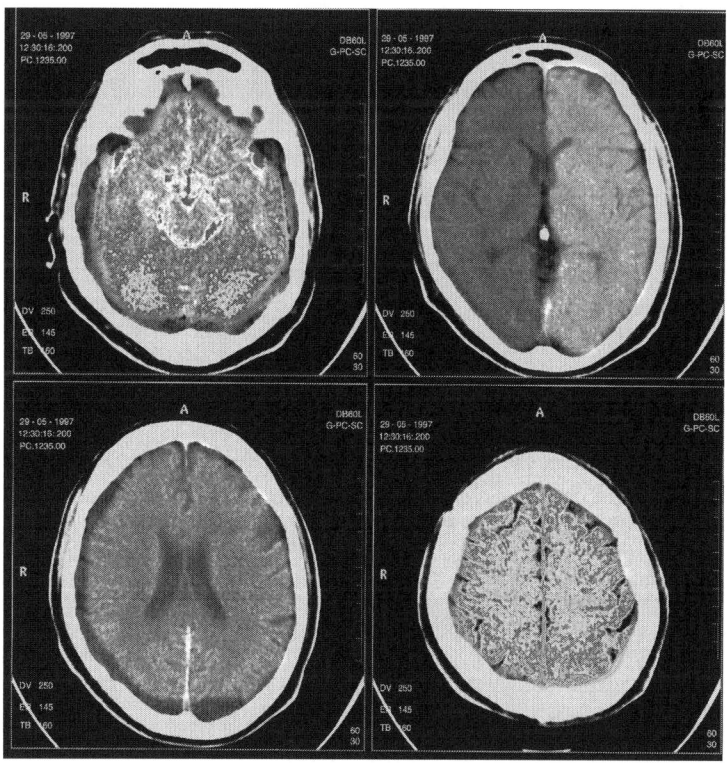

CAT scanning of the head is typically used to detect tumors, calcifications, hemorrhage, and bone trauma. (iStockphoto)

only choices would be a "0" or a "1." A two-bit system would gives four choices per pixel, "00" (black), "01" (dark grey), "10" (light grey), and "11" (white). Typically, in photo editing programs of the early twenty-first century, each pixel is described by 24 bits of information, yielding more than 16 million possible colors.

Resolution

Bitmap images contain information for a given number of pixels. The larger the pixel number, the more information is in the image and the higher the resolution; typically, this also results in a bigger file. Screens are all made of pixels, whether they are on computers or cell phones; if an image is viewed at full size, each pixel in the image will show up as one pixel on the screen. However, if a viewer zooms in beyond this point, the pixels in the file are actually represented by big blocks of pixels on the screen, and the image is said to become "pixelated."

Thus, if an image is to be viewed on a screen, it will ideally have the same number of pixels as the size one wants it on the screen; any more than that is wasted file space, and any fewer will result in an image that appears pixelated. If images are going to be printed, however, more pixels will translate into sharper pictures, limited only by the resolution of the printer. Again, the larger the print, the more pixels you will need for a sharp print.

File Types and Compression

Bitmap graphics can be stored in a variety of file formats depending on how they will be used. Raw files, which store all the raw data for the light that hits each CCD pixel, are commonly used by photographers who wish to have maximum flexibility and are not worried about file size. In order to make files smaller, computers use mathematical algorithms to compress the files. For example, instead of recording values for each pixel, the values for some could be calculated by the difference between a pixel and its surroundings, thus yielding substantial file size savings where blocks of pixels are the same as their neighbors. Some kinds of compression are considered "lossless," because all the information from the original can be re-created when the file is decompressed. However, there are a number of compression schemes such as the popular jpeg format in which the mathematical approximations do not quite match the original. In these cases, accuracy is sacrificed in order to save file size, and these approaches are said to be "lossy." However, the algorithms used to compress and decompress files are generally so good as to be unnoticeable in many cases. The JPEG 2000 image compression standard for both lossless and lossy compression uses biorthogonal wavelets, which extends from the work of mathematician Ingrid Daubechies, known as the "mother of wavelets."

Vector Graphics

Certain kinds of images, especially those created in computer graphics programs, use a different method for describing the content of the image. Instead of denoting each pixel with a number, these vector graphics are described mathematically as a set of equations representing the lines and curves that make them up. When a viewer zooms in on a vector graphic, the image does not become pixelated, because the computer recalculates the curve or line based on the new image size. While vector graphics are not appropriate for photographs, photo editing programs may use them when overlaying text or graphics on a digital image.

Image Reconstruction

The basic problem of image reconstruction is to build a "best-guess" object out of averaged data and then estimate how close the reconstruction is to the actual object. For example, in a single-angle X-ray of a person, the amount of radiation going in and coming out the other side can be measured and visualized on X-ray film. The difference between the values is how much was absorbed, but there is limited information about the inner structures that blocked the radiation. This limitation can make diagnoses difficult. However, if the same person is X-rayed from several directions and angles, the resulting information can be compiled, averaged, or mathematically modeled to estimate what the internal structure looks like.

Further Reading

Alsina, Claudi. *Math Made Visual: Creating Images for Understanding Mathematics*. Washington, DC: The Mathematical Association of America, 2006.

Hoggar, S. G. *Mathematics of Digital Images: Creation, Compression, Restoration, Recognition*. Cambridge, UK: Cambridge University Press, 2006.

JEFF GOODMAN

Digital Storage

Category: Communication and Computers.
Fields of Study: Algebra; Measurement; Number and Operations.
Summary: Information can be stored digitally—a process that requires information to be translated into binary code.

Digital information is information in binary code. In order to create, manipulate, and store this digital code, it must be created in physical form. This creation is done by using media that can exist in one of two distinct states and assigning one state to each of the two digits ("0" and "1") in binary code. Within a computer, the "1"s and "0"s are represented as "ons" and "offs"; on a magnetic hard disk, they are tiny magnets pointing one way or another; and on a CD, the two states are shiny and dull spots. Engineers used metal tape on reel-to-reel machinery to record audio signals in the early twentieth century. In 1952, IBM introduced a tape drive with iron oxide–coated plastic tape. Reel-to-reel tape drives were the standard for data storage by the mid-1970s. IBM also created magnetic hard disks in the late 1950s, but it took decades to overcome size and access speed issues to make hard disk drives (HDDs) feasible for applications like personal computers. Solid-state drive (SSD) technology, such as flash memory, was the necessary next step to overcoming the lagging mechanical speeds of HDDs. Mathematicians in many fields have been essential in all stages of development and continue to address emerging issues. Ingrid Daubechies, "the mother of wavelets," is perhaps best known for her work with wavelet-based algorithms for compressing digital images. Irving Reed and Gustave Solomon developed algebraic error-detecting and error-correcting codes. These Reed–Solomon codes are widely used in digital storage and communication, from satellites to CDs.

Bits and Bytes

The smallest unit of stored digital information, corresponding to a single "1" or "0," is called a "bit." The term "bit," a contraction of "binary digit," is commonly attributed to statistician John Tukey, working in conjunction with mathematician John von Neumann. Bits are collected into 8-unit chunks called "bytes," and these collections of 8 bits can represent various types of information. The lowercase letter "a," for instance, can be represented as 01100001, and "b" as 01100010. The music on a compact disc is encoded as a set of 44,100 reading (or samples) per second, with each reading represented by 2 bytes containing 16 bits.

Storage Size

Sizes of files, and the capacity of storage devices, are often referred to as multiples of the byte. A kilobyte (KB) is approximately 1000 bytes, enough information to store about 150 words, or about half a page of text from a paperback book. As larger units are used, the naming system employs other metric prefixes, with each step up representing a multiple of either 1000 or 1024, depending on the device. Thus, a megabyte (MB) is approximately 1000 KB, and a gigabyte (GB) is approximately 1000 MB. Units beyond the gigabyte include the terabyte (TB), petabyte (PB), and exabyte (EB).

Magnetic Storage

Since grains in a magnetic medium can be magnetized with the north pole pointed in either of two directions, magnetism is an ideal medium for representing binary information. In addition, since information stored in this way is relatively stable, it is useful for long-term storage. Finally, since this magnetism can be reset easily using an electromagnet, magnetic media are easy to erase and rewrite.

A magnetic hard disk employs one or more spinning platters coated in a magnetic medium. An arm with tiny electromagnetic heads floats over the surface of the disk and is used to magnetize regions of the disk corresponding to the "1"s and "0"s of binary code. To retrieve information, the disk spins past the heads, generating current that corresponds to the code stored on the disk. While the principle is straightforward, it has been a remarkable feat of engineering to create disks that spin up to 7200 revolutions per minute with arms that can travel across the surface of a platter 50 or more times per second as they seek and write information. Even so, writing and retrieval speeds have not increased over time at the same exponential rate as the amount of information that can be stored on such disks, resulting in undesirable lags.

Even in the early twenty-first century, long-term backup of computer information is often done on low-cost magnetic tape, with bits of information laid down as magnetic regions on moving tape. However, since the information is laid down on a long piece of tape, there can be no random access of information, limiting its usefulness in everyday applications. Until recently, digital camcorders used magnetic tape to record video; however, the desire to have random access of footage and recent advances in hard drive and other storage techniques have brought on a new generation of tapeless camcorders.

CDs, DVDs, and Flash Memory

Both CD and DVD players are optical devices that use lasers to read the shiny and dull spots encoded on a plastic disk. Information is recorded by burning non-reflective pits into the surface of the disk to represent "0"s and leaving the reflective surface to represent "1"s. When the disk is played, it spins past a laser. When the light encounters a pit, it is not reflected, and the player registers an "off" signal ("0"), and when the light bounces back off a shiny region, the player registers an "on" signal ("1"). This information is interpreted by a small computer in the player.

Many devices, including digital cameras, camcorders, video game consoles, and cell phones, use flash memory, which can store large amounts of information on small cards that have no moving parts. This technology employs an array of microscopic transistors through which current may pass. Whether this current passes through or not is controlled by what is called a "floating gate," and the path through the transistor can be electrically opened or closed. This method allows the transistor to have the two states needed for binary code. Sections of flash memory can easily be reset (erased) by flushing out the electrons trapped in the floating gate. One of the primary benefits of this technology is that information can be stored on a card with no moving parts, improving both access speed and portability.

Data Rot and Error Correction

Tape, hard disks, CDs, and flash memory store and retrieve information accurately most of the time, but they are not problem-free. Errors and noise can happen in an electromechanical recording system—"1"s that should have been "0"s, and vice versa—which diminish information accuracy. Mathematical methods are used to check for and correct errors. For example, cyclic redundancy check (CRC) coding algorithms calculate a fixed-length binary sequence (code) for each block of data using polynomial division in a finite field. The codes and data blocks are stored together, and they can be checked after transmission or retrieval. CRC was invented by mathematician W. Wesley Peterson, who also devised many error-correcting codes.

Even if the recording is perfect, the media that hold binary code can degrade in a variety of ways over time. For instance, magnetic media can lose their magnetic orientation, especially if they are subjected to a strong magnetic field. In addition, the substrates on which the magnetism is stored—the platters on hard drives and plastic backing on magnetic tape—will invariably degrade over time. Even the plastic on CDs and DVDs will begin to break down, and flash memory floating gates will ultimately leak the electrons that maintain data in their flash memory transistor states. Even if the storage media and binary information survive over time, there is a real chance that in the future there may not be hardware available to read information encoded in an outdated media.

Further Reading

Somasundaram, G., and Alok Shrivastava. *Information Storage and Management*. Hoboken, NJ: Wiley, 2009.

Wicker, Stephen. *Error Control Systems for Digital Communication and Storage*. Englewood Cliffs, NJ: Prentice Hall, 1994.

Jeff Goodman

Electricity

Category: Architecture and Engineering.
Fields of Study: Algebra; Representations.
Summary: Electricity, arising from the flow of electrons, can be described mathematically.

Daily operations of modern industrial societies, including transportation, communication, heating, cooling, lighting, computing, and medical technology, rely on the use of electrical power. Power from batteries and electrical outlets is derived from the flow of electrons, known as "electric current." The term "electricity" refers to a variety of physical effects, both static and dynamic, that arise from electric charge. The mathematical description of electric and magnetic phenomena developed in the eighteenth and nineteenth centuries contributed to a rapid expansion of electrical technology, which is powered today by a vast grid of electric power stations and distribution systems.

Electric Charge and Coulomb's Law

Electric charge is a property of matter that can be negative (as in electrons), positive (as in protons), or zero. Most matter has a net charge of zero, containing essentially the same number of electrons as protons. Two objects whose charges are both positive or both negative repel each other, while objects with opposite charges attract each other. Static electricity is created when electrons build up on or are depleted from the surface of a material, often by rubbing materials together. Effects of static electricity are seen, for example, in a rubbed balloon clinging to a wall, or in hair standing on end. In metals, electrons are not strongly bound to individual atoms but move freely through the lattice of protons. Materials with freely moving charges are known as "conductors." The force between two charged particles at rest is described by Coulomb's Law, named after French engineer Charles-Augustin de Coulomb (1736–1806). Coulomb's Law states that the magnitude F of the force exerted by one charged particle on the other is

$$F = \frac{kqq'}{r^2}$$

where q and q' are the magnitudes of the charges of the particles, r is the distance between the two particles, and k is a constant. This equation shows, for example, that if one charge is tripled, then the force is tripled, and if both charges are tripled, then the force becomes nine times as large. On the other hand, tripling the distance r between the particles multiplies the right-hand side of the equation by $1/3^2$, or $1/9$, reducing the force to a ninth of its previous value.

Electric Field and Electric Current

The presence of charged particles creates an electric field that exerts a force on other charged particles in the region. An electric power generator, usually driven by a steam turbine fueled by coal or a nuclear reactor, creates an electric field between two terminals by building an over-supply of electrons (negative charge) in one terminal and a deficit of electrons (positive charge) in the other. The flow of electrons from a negative toward a positive terminal along a conducting path, such as a wire, is an electric current. In lightning, electrons from negatively charged clouds in the atmosphere are attracted to positively charged objects on the ground beneath the cloud. Here the electric field is so strong that electric current passes through air, which usually acts as an insulator that prevents the flow of electrons. Batteries operate by producing an electric current between oppositely charged terminals of chemical cells. A battery produces direct current (DC), where electrons flow in one direction, while a power generator creates alternating current (AC), where the direction of electron flow alternates rapidly, typically at a frequency of 60 hertz (cycles per second). The hertz is named for German physicist Heinrich Hertz (1857–1894), who made important advances in understanding the connection between electric and magnetic fields.

Ohm's Law

The energy that an electric field imparts to a unit charge moving from one terminal to another is the number of volts (V) between the terminals, named after Italian physicist Alessandro Volta (1745–1827). On electric bills, energy usage is typically given in kilowatt hours (kWh). The watt, named for British engineer James Watt (1736–1819), is a unit of power, or energy per time, and 1 kilowatt is 1000 watts. Multiplying power (in kilowatts) by time (in hours) yields energy, in kilowatt-hours. In an electric current, the current intensity (I) is abbreviated as "current" and is the quantity of charge that moves past a cross-section of the conducting path per unit time. As electric current flows through a material, the motion of the electrons is hindered by positive ions, creating electrical resistance (R). Resistance in the path of a current creates heat and light, as in appliances, such as stoves and light bulbs. Electrical energy can be transformed into mechanical energy to power motors as in cars, airplanes, power tools, kitchen blenders, and hair dryers when electric current passes through a coil of wire, inducing a magnetic field that sets the coil in motion.

Ohms's Law, formulated by German physicist Georg Ohm (1789–1854), states that for a metal conductor at constant temperature, the voltage (V) is $V = IR$, where I is the current, and R is the resistance. This equation shows, for example, that if the resistance is cut in half, then to maintain the same voltage, the current must be doubled. If too little resistance is present, the current may become so strong as to damage electrical equipment. Circuit breakers then sever the path of the current to avoid damage.

Electric Power from Generator to Consumer

High voltage generated at power stations is propagated along power lines almost instantaneously, over many miles, to substations near cities and towns. At the substations, the voltage is reduced and transmitted to electric distribution centers that channel the voltage to homes, offices, and other facilities. In standard electrical outlets in the United States, there are 120 volts between the wires leading to the two vertical slots. When an appliance is plugged into the outlet, the vertical prongs of the plug make contact with these wires, creating a pathway of current through the appliance. The third slot in the outlet carries a protective ground wire. In appliances with a three-pronged plug, the ground wiring is designed to provide a preferred pathway for escaped current so that it will not travel through the body of the person holding the appliance.

Large appliances, including most drying machines and ovens, operate at 240 volts, using a different type of outlet. Touching one or more openings in an electrical outlet or touching the prongs of a plug as it is inserted into the outlet may pass an electric current through the body that can be harmful or even deadly. At electrical facilities, "High Voltage" signs warn of the danger of electric shock because of the presence of high voltage.

Further Reading

California Energy Commission. "What Is Electricity?" http://energyquest.ca.gov/story/chapter02.html.

Herman, Stephen L., and Crawford G. Garrard. *Practical Problems in Mathematics for Electricians*. 6th ed. Albany, NY: Delmar, 2002.

U.S. Energy Information Administration: Independent Statistics and Analysis. "Electricity." http://www.eia.doe.gov/fuelelectric.html.

Barbara A. Shipman

Elevators

Category: Architecture and Engineering.
Fields of Study: Algebra; Number and Operations.
Summary: Mathematics is used to quantify aspects such as the maximum speed and distance range of elevators as well as model vibration and optimize traffic flow.

An elevator is a mechanism for vertical transport of persons or cargo. Mathematics is used to quantify aspects such as the maximum speed and distance range of elevators, determined by their purpose, such as lifting passengers, cars, or aircraft. Applied mathematical models focus on the dynamics and vibrations within different types of elevator mechanisms, such as hydraulic or rope systems. Mathematicians also investigate questions related to aspects such as waiting time, using probability models. Systems of multiple elevators are modeled as high-dimensional spaces using dynamical systems. The number of passengers in an elevator system constantly changes, making an optimal policy for what is referred to as an "elevator group control" mathematically interesting. At the end of the nineteenth century, scientist Konstantin Tsiolkovsky conceived of a space elevator. He was self-taught and worked as a mathematics teacher.

Hydraulic Elevators

The main concept related to why hydraulic elevators work is Pascal's Law, stating that when the pressure increases anywhere in a confined fluid, it equally increases everywhere. This, together with the fact that pressure (P) is equal to force (F) per unit area (A), can be exploited for an advantage of force. The elevator car stands on top of a piston ending in a wide shaft filled with oil, connected to a narrow shaft with oil. When a pump increases pressure in the narrow shaft, by applying a relatively small force, the equal pressure applies to the floor of the cabin, producing higher force because of the larger area: $P_1 = P_2$, and

$$\frac{F_1}{A_1} = \frac{F_2}{A_2}.$$

Hydraulic elevators are only used in relatively low buildings since the piston has to be as tall as the building to extend to the top floor but fully fit under the building when the elevator is on the ground floor. Digging as deep as a skyscraper is high to install an elevator is impractical. These elevators are mostly used for heavy loads in places such as car mechanic shops.

Roped Elevators

A mathematically interesting concept related to roped elevators is the conservation of energy. A roped elevator consists of two ends of a steel cable going around a pulley attached at the top, called a "sheave." The elevator car is attached to one end of the cable, and the counterweight, which weighs about the same, is attached to the other end.

When the elevator car is at the bottom of the shaft, the counterweight is at the top, and its potential energy converts to force, helping move the elevator car up. When the elevator car is higher than the counterweight, their roles are reversed. This way, it takes very little additional force to make the sheave rotate and the elevator car move up and down.

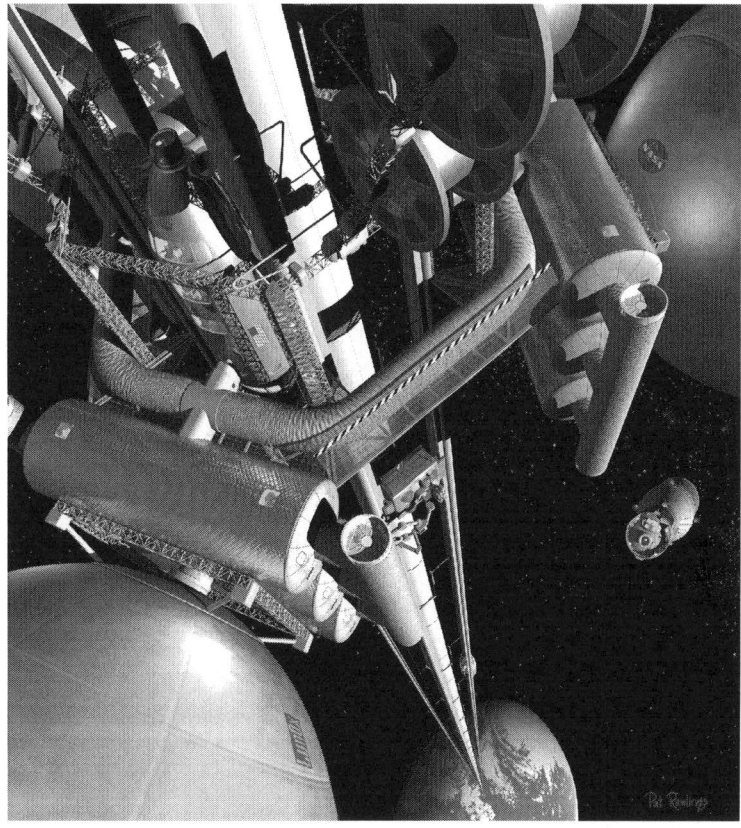

The National Aeronautics and Space Administration (NASA) holds an annual engineering competition to design a space elevator. (National Aeronautics and Space Administration)

Logistics

In modern buildings with multiple elevators, computer programs determine how to dispatch elevators to minimize wait time and to save energy. For example, a sensor may detect that an elevator is near capacity and will not stop it for any additional passengers. An elevator going down may not open its doors for people who want to go up, avoiding carrying them back and forth. More sophisticated elevator software can take into account typical traffic patterns, directing elevators to the busiest floors.

Space Elevator

A space elevator is a structure for escaping the gravity well of a planet, transporting objects between the surface and a geostationary orbit. This proposed structure would consist of a large satellite counterweight in orbit and a cable connecting it to the ground. The inertia of the counterweight rotating around the planet will balance the gravitational pull on the cable, keeping the cable taut. The National Aeronautics and Space Administration (NASA) is working on several efforts related to construction of a space elevator, including an annual engineering competition. The technological problems include avoiding meteorites and dangerous atmospheric weather systems, developing materials strong enough for the cable, designing the counterweight, protecting passengers from radiation, and powering the elevator cars. In 2008, Japan announced plans to build a space elevator in the immediate future. Space elevators have frequently appeared in science fiction since the early twentieth century.

Further Reading

Bangash, M. Y. H., and T. Bangash. *Lifts, Elevators, Escalators and Moving Walkways/Travelators*. Leiden, The Netherlands: Taylor and Francis, 2007.

Van Pelt, Michel. *Space Tethers and Space Elevators*. New York: Copernicus Books, 2009.

Wuffle, A. "The Pure Theory of Elevators." *Mathematics Magazine* 55, no. 1 (January 1982).

Maria Droujkova

Engineering Design

Category: Architecture and Engineering.
Fields of Study: Algebra; Geometry; Measurement.
Summary: Engineering design is a carefully regulated process to create optimal solutions for given problems.

Engineers design everything from automobiles and bridges to prosthetic limbs and sporting equipment. Designing is different than simply building in that it requires the adherence to a very systematic, yet iterative, process known as the "engineering design process." This process is to engineers what the scientific method is to scientists—guiding steps that help ensure that the end result is the best it can be. When a new product is created without following the steps of the engineering design process, there is a higher likelihood that the product designed will lack some important aspect: the end product may not appropriately account for the needs of its users, it may cost too much to manufacture, or it may not have been tested to ensure safety. Accordingly, the term "designing" refers to the entire process, such that an engineer "does design." The use of the term "design" as a noun may be used at different points in the process but may have very different meanings depending on what phase of the process the engineer is in. Design may really mean "design idea" during the brainstorming phase of the process or "model or prototype of the design" during the building phase of the process.

The engineering design process requires the application of mathematics in many of the steps. Throughout the process, engineers use basic mathematics concepts, including addition and multiplication to calculate costs; geometry to calculate surface areas for material needs; and measurements to ensure appropriate dimensioning. However, more sophisticated projects may require the application of higher-level mathematics, such as calculus and differential equations, to solve the technical engineering problems certain designs pose.

The Engineering Design Process

The engineering design process refers to the steps that are required to create the best possible solution to a problem. It is a process often undertaken by a team of engineers who work together, though it can be performed by an individual—trained or untrained as an engineer. Though there is no consensus as the exact breakdown and name of each step, the general design process is universally accepted.

In the first step of the engineering design process, the engineering team is presented with some type of problem or unmet societal need to be solved. Often, this problem is presented to the engineering team by a company that is trying to offer a product that better meets its customers' needs. The engineer must ask many questions to both the client and the user, as well as conduct background research, in an effort to establish the objectives and constraints of the design. The objectives are what the solution to the problem (the final designed product) should aim to accomplish. The constraints are the factors that limit the possible designs, such as time, money, or material restrictions. Time and money constraints are particularly important as they often drive the project and must be monitored throughout to ensure that the project is completed on time and within budget. At the end of this step of the design process, the engineering team fully understands the problem and has developed objectives and constraints to guide their possible solutions.

In the next step of the engineering design process, the engineers generate design ideas to solve the newly refined problem. Idea generation normally occurs through group brainstorming methods, with the goal of producing as many ideas as possible. There are a number of methods used to enhance the innovation and creativity of the ideas that come from the brainstorming session, including ensuring group diversity, drawing from existing stimulus and building off of each other's ideas. In this step of the process, some of the generated ideas will evolve into rough hand-drawn sketches. These sketches need to show perspective and relative size clearly.

The next step of the engineering design process is design selection. A method known as "decision analysis" is most commonly used for design selection. Decision analysis is a systematic process to objectively and logically choose the best idea to move forward with from the many generated through brainstorming. It is important because it reduces the likelihood of a designer's bias in selecting a design. As a first step, the brainstormed ideas must initially be narrowed down through discussion or other means to only the handful of ideas that appear to be most promising. These ideas are then compared through decision analysis. For the decision analysis, it is first necessary to create a list of design criteria and weight them based on their relative importance. As an example, as safety is paramount in design, the criteria of "safety" would be the most important criteria and would be weighted as 1.0 on a scale of 0–1. The criteria of "portability," on the other hand, might be desirable but not necessary, so it would be weighted as 0.5. There is no standard as to what weighting scale should be used but it is important to be consistent in its application. For each criterion, in addition to the determined weighted importance, a numerical range must also be established for rating each design with respect to the criterion. When possible, this range should be as objective and quantifiable as possible.

Each design being considered is then "scored" using the range for each criterion. The score is then multiplied by the relative criteria weight for a total score for each criterion and for each design. The total scores for each criterion are then summed for each design. The summed scores can be used to compare multiple designs, with the one scoring the highest being the one most likely to be successful.

After identifying a design to move forward with, refinement of the design is necessary. This step includes determining dimensions and materials that will be used to construct the chosen design. Detailed sketches, often drawn from multiple perspectives, are created and include the dimensions of each part to be made. Determining these dimensions often requires in-depth estimation and calculation. At the most simplistic level, dimensioning requires taking into account any necessary clearances or gaps in the design, especially when multiple parts need to be fitted together. It may also be necessary to determine the combinations of dimensions that ensure a specified surface area requirement is met, in which case algebra can be helpful. More in-depth designs may require that dimensions come from established tables of normative dimensions, such as anthropometric tables, providing typical measurements of different-sized people, or from engineering analysis, such as stress or buckling calculations. Deriving dimensions from engineering analysis methods often requires high-level mathematics and a technical background in engineering but ensures a stronger, safer product.

Once the design has been refined and the dimensions are known, building begins. For most designs, a scale model or a simplified prototype is created first to test for feasibility of the design before further time and money is invested. To create a scale model, all dimensions of the detailed sketches must be reduced by multiplying by some chosen scaling factor, often 1:2. Regardless of whether a full-size design or scale model is used, it is necessary to calculate the amount of each material that needs to be purchased to build the design. This requires thought and calculation, in particular when multiple parts could be cut from one piece of wood, metal, or fabric. Often, surface area is calculated according to the part's geometry to determine the total amount of material needed. Once material has been secured, building of the design can occur. Throughout building, it is essential to make careful measurements for all parts because

Once the engineering team is satisfied with the final product, the design is executed through computer-aided design (CAD) drawings. (iStockphoto)

almost all designs are made from multiple components that must fit together to function as one product. For example, if a piece of wood to be used for one leg of a chair is measured even ¼ inch shorter than the other legs, it will likely mean the finished chair will rock and wobble, and the design will be undesirable.

As a next step in the engineering design process, the constructed design is experimentally tested to determine its performance. This step helps to identify design strengths and weaknesses, which can be used to make recommendations for future refinement of the product. The specific experimental test performed is determined by the type of product designed and the design objectives. Regardless of the type of test conducted, measurements are taken throughout the experiment to record some aspect of the design's performance. Often, multiple trials will be taken, generating many data points. The data obtained from these measurements are then used to draw conclusions about the success of the design. Statistical analysis may also be employed to further assist in the interpretation of the data.

Almost always, the data collected during testing will suggest that the design could perform better if refined in some way. As such, it is common for the engineering team to return to the building stage and then iteratively cycle between it and testing steps until satisfied. At times, it may also be necessary to return to earlier steps in the engineering design process. Once the team is satisfied with the final product, final documentation is prepared to explain the design and share it with others. This is often done through computer-aided design (CAD) drawings and written technical reports.

Further Reading

Dym, Clive L. and Patrick Little. *Engineering Design: A Project Based Introduction.* Hoboken, NJ: John Wiley and Sons, 2009.

Eide, Arvide, et al. *Introduction to Engineering Design and Problem Solving.* New York: McGraw-Hill, 2001.

Pahl, Gerhard, et al. *Engineering Design: A Systematic Approach.* 3rd ed. New York: Springer, 2007.

Pilloton, Emily. *Design Revolution: 100 Products That Empower People.* Los Angeles, CA: Metropolis Books, 2009.

Ulrich, Karl, and Steven Eppinger. *Product Design and Development.* New York: McGraw-Hill/Irwin, 2007.

Kimberly Edginton Bigelow

Fax Machines

Category: Communication and Computers.
Fields of Study: Number and Operations; Representations.
Summary: Fax machines revolutionized the process of sending and receiving documents.

A fax machine enables documents, including illustrations and other graphical elements, to be transmitted over a distance and reproduced by the receiver. The roots of the word "facsimile" are from the Latin words *facere*, meaning "to make," and *similis*, meaning "like." In the nineteenth century, Alexander Bain developed what some refer to as the first fax machine. His system transmitted information using analog telegraph lines. The sending and receiving equipment was timed using matched pendulums. At the receiving end, an electrically powered stylus recorded messages on a roll of paper. Current from the stylus turned the chemical coating on the paper blue, transcribing the signals' dots and dashes. Frederick Bakewell demonstrated a chemical fax machine at the 1851 London Exhibition, and the first commercial telefax service began operation in 1865, predating the telephone.

A more modern ancestor is the radio facsimile, developed in 1924, which used radio waves to wirelessly transmit images and is still used in the early twenty-first century to transmit weather information. Modern fax machines scan an input sheet line by line to produce rows of pixels. Algorithms used in fax machines take advantage of the fact that there are white and black pixels in order to compress the data. For example, David Huffman's variable-length lossless codes and their variations, originally invented in the 1950s, assign binary codes to patterns of pixels using probabilistic methods. The codes are shorter than the strings they replace, reducing overall file size. To optimize compression, symbols with higher probabilities or frequencies of occurrence are assigned shorter codes. The International Telecommunications Union, based in Geneva, Switzerland, makes recommendations for data compression standards. To derive one code called the "Group 3 code," the organization applied the Huffman algorithm to eight representative samples to assign a code to each run length. Fax machines transmit documents in minutes instead of hours thanks to compression algorithms.

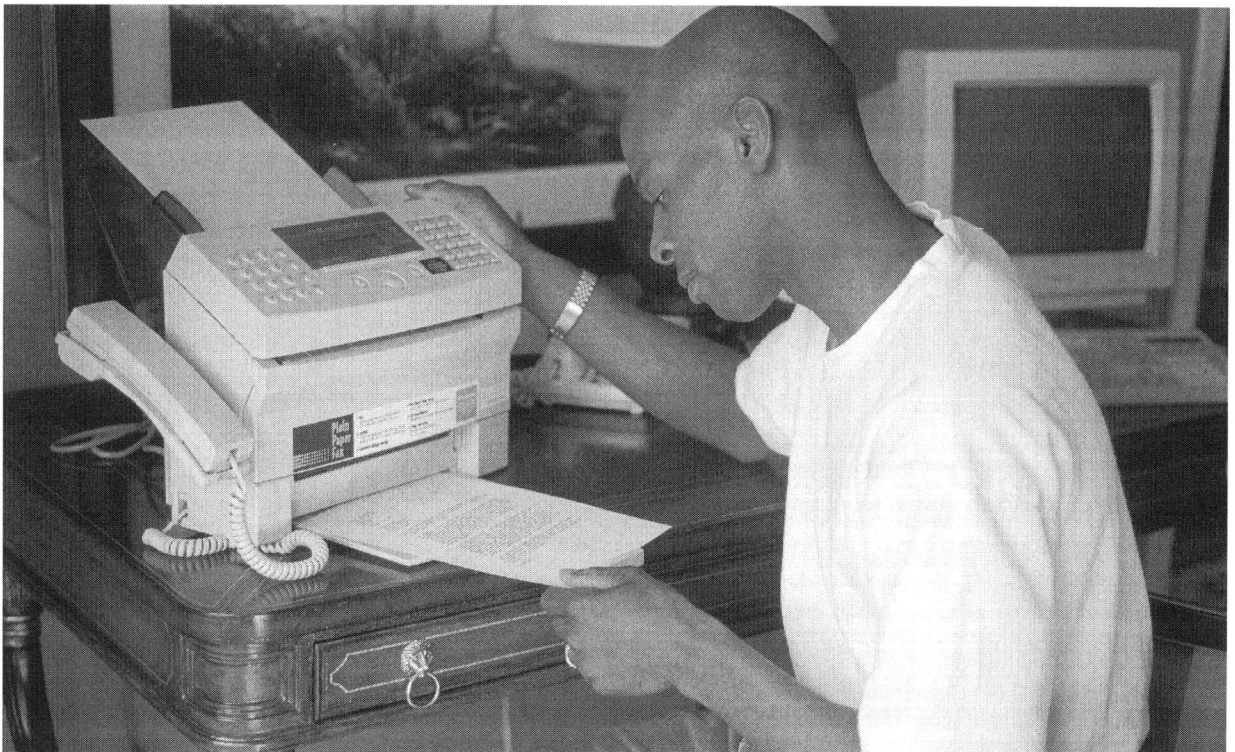

The sending fax machine uses a sensor to scan the document and to convert the pattern of black and white elements into a code. The receiving machine decodes and uncompresses the information and prints it. (iStockphoto)

Fax Machine Technology

Modern fax machines utilize the technology of the telephone and the copy machine. Fax machines developed in the 1970s could scan a document and encode and transmit it over telephone lines to another fax machine, which could record and reproduce the document. Fax machines became common in offices as they replaced the need to send paper documents by messenger service or mail, were much quicker than retyping a document for telex, and could send any type of graphical information. Japan played an important role in developing modern fax machines, which used electronic circuits to replace mechanical parts and greatly increased the speed of transmission and reduced the size and price of the machine. Because the Japanese language incorporates many Chinese characters (*kanji*), the ability to transmit graphical images was particularly useful in that country.

Sending a document by fax requires two fax machines—one to send the document and one to receive it. The sending machine uses a sensor to scan the document, usually line by line, and to convert the pattern of black and white elements on the page into a code (several coding standards exist). The fax machine is not "reading" text—in the sense of converting the letters into meaning—but only recording their shape. For this reason, fax machines are as adept at sending images and diagrams as they are at sending text. The scanned data are compressed in order to reduce the number of bits to be transmitted and thus to speed up the process. The speed of transmission depends in part on how much information, such as text or diagrams, as opposed to blank space is contained on the page being scanned. The receiving fax machine decodes and uncompresses the information and uses it to re-create and to print the sent document. In the 1980s, most fax machines used thermal printing, which required the use of special paper that turns black when exposed to heat. However, in the twenty-first century, most fax machines print on standard white copy paper using either laser or inkjet printing technology.

Internet fax (efax, or online fax) technology has supplemented and, in some cases replaced, the use of traditional fax machines. There are a number of different services offering Internet fax capability, and although they differ in some details (for instance, can the machine receive, send, or both) the principle is the same: they provide a means to transmit facsimile documents to and from computers either as e-mail attachments or through a dedicated phone number or Internet site.

Further Reading

Brain, Marshall. "How Fax Machines Work." http://communication.howstuffworks.com/fax-machine.htm.

McConnell, Kenneth R., Dennis Bodson, and Stephen Urban. *FAX: Facsimile Technology and Systems*. 3rd ed. Boston: Artech House, 1999.

Salomon, David. *A Guide to Data Compression Methods*. New York: Springer, 2002.

Smithsonian Institution. "From Carbons to Computers: The Changing American Office." http://www.smithsonianeducation.org/scitech/carbons/start.html.

Sarah Boslaugh

Fuel Consumption

Category: Travel and Transportation.
Fields of Study: Algebra; Measurement.
Summary: Vehicle fuel consumption and efficiency are often mathematically investigated.

Fuel consumption can be defined as the amount of fuel used for each unit of measurement (usually time or distance). An often mistaken meaning is fuel economy, which is the reciprocal of fuel consumption: the amount of distance or time for each unit of fuel used. In addition to people in all walks of life using mathematics to measure fuel consumption, mathematicians research fuel optimization.

In one case, Pontryagin's Maximum Principle, named after mathematician Lev Pontryagin, characterizes optimum values that determine a trajectory, such as fuel consumption or flight time. On the other hand, the counterintuitive assertion that greater fuel efficiency often results in increased fuel consumption is sometimes known as the Jevons Paradox, after economist and logician William Jevons. Mathematicians are also involved in research for alternative fuel sources for vehicles, such as biodiesel and electrical power.

Fuel consumption is calculated for various reasons, including budgeting and maintenance. If a business that uses fuel knows the average amount of time or average distance traveled by its machines and the fuel consumption for each unit, it can calculate the approximate amount of money needed for purchasing fuel over the next fiscal period. Tracking fuel consumption on a regular basis can indicate a potential breakdown of internal engine parts before the issue becomes a major repair or hazardous situation.

Calculating Fuel Consumption

Calculating fuel consumption is a fairly simple process if you have a way to measure both the time or distance the machine was used and the amount of fuel used to refuel the machine. For example, machines that are designed for travel, like cars, trucks, vans, or tractor-trailers contain an odometer to record the number of miles or kilometers traveled. Many even have a trip odometer that can be reset after refueling. To calculate fuel consumption, start by having the vehicle completely filled with fuel and the odometer reading recorded or reset.

After using the vehicle, fill its tank with fuel and measure the amount of fuel that has been added. The assumption here is that the amount of fuel added to bring the tank back to its full position would approximate the amount of fuel used since the last time the vehicle was fueled. At the same time that the vehicle is refueled, also record the odometer. A trip odometer indicates the exact distance traveled since the last fill-up (the distance traveled since it was last reset). If not using a trip odometer, take the current total distance traveled and subtract the previous reading taken at the last fill-up.

Now that the distance and amount of fuel has been measured, calculating fuel consumption is the simple division problem $C = F \div D$, where C is the fuel consumption, F is the amount of fuel used, and D is the distance traveled.

For example, a vehicle that traveled 400 miles on 20 gallons of fuel has a fuel consumption of $20 \div 400 = 0.05$ gallons per mile, meaning that five-hundredths of a gallon (6.4 fluid ounces) of fuel was used to travel each mile. In Europe, Australia, and other countries (like Canada and China) fuel consumption is calculated in liters per 100 kilometers traveled.

A vehicle that traveled 600 kilometers on 75 liters of fuel would have a fuel consumption of $75 \div 600 = 0.125$ liters per kilometer. To get liters per 100 kilometers, multiply the result by 100 to get 12.5 liters per 100 kilometers. When looking at fuel consumption, a lower number is better than a higher number, meaning you use less fuel to achieve the same distance.

Some countries use fuel economy; for example, the United States uses miles per gallon and Japan uses kilometers per liter. The formula for fuel economy (E) is $E = D \div F$.

In the above examples, $400 \div 20 = 20$ miles per gallon (mpg), and $600 \div 75 = 8$ kilometers per liter. For fuel economy, a larger number is desired, meaning a greater distance can be traveled using the same amount of fuel.

Not all machines were designed to travel, such as forklifts and construction equipment. Generally, these machines do not measure the distance they have traveled but rather the number of hours the machine has been in use. Many of these machines have an "hour meter" that measures the time the machine operates. For example, if a forklift uses 5 gallons of fuel over an 8-hour shift, fuel consumption is found by the formula $C = F \div T$ where T is the time the machine is in use. In the above example, $5 \div 8 = 0.625$ gallons per hour.

Further Reading

O'Hayre, Ryan, Suk-Won Cha, Whitney Colella, and Fritz B. Prinz. *Fuel Cell Fundamentals*. Hoboken, NJ: Wiley, 2009.

Ross, Michael. *A Primer on Pontryagin's Principle in Optimal Control*. Carmel, CA: Collegiate Publishers, 2009.

U.S. Department of Energy. "Save Money." http://www.fueleconomy.gov/feg/savemoney.shtml.

Woodsie, Christine. *Energy Independence: Your Everyday Guide to Reducing Fuel Consumption*. Guilford, CT: Lyons Press, 2009.

CHAD T. LOWER

GPS

Category: Travel and Transportation.
Fields of Study: Geometry; Measurement; Number and Operations.
Summary: Global positioning systems have been made available to the private sector but depend on satellites originally placed into orbit for military purposes and require precise calculations.

The global positioning system (GPS) is a satellite-based navigation system comprised of a network of satellites placed into orbit by the U.S. Department of Defense in 1973. GPS was originally intended for military applications to accurately determine locations worldwide in all kinds of weather. In the 1980s, the U.S. government made the system available for civilian use. GPS is used as a navigation and positioning tool in transportation, such as fleet cars and commercial trucking, in surveying, and for almost all outdoor recreational activities. In the scientific community, GPS plays an important role in geology, meteorology, wildlife studies, archeology, and many other areas. Mathematics was critical in the development of this system and mathematicians work on many ongoing issues, such as precision and error correction.

There are three parts that form the GPS: the space segment (satellites), the user segment (the receiver), and the control segment (control stations). The control segments are on the geoid (a three-dimensional model of Earth). The first segment of the system consists of a constellation of satellites, orbiting 20,000 kilometers above Earth in 12-hour circular orbits. While the exact number of satellites in operation varies at any given moment, at least six groups of four satellites are necessary to ensure that they can be detected from anywhere on Earth's surface. Each group is assigned a different path, creating six orbital planes that completely surround Earth.

Trilateration

The satellites transmit signal information to Earth. GPS receivers take this information and use trilateration to calculate the user's exact location. Each satellite continuously transmits a data stream containing orbit information, equipment status, and the exact time. GPS receivers contain computer chips that then calculate the difference between the time a satellite sends a signal and the time it is received. The unit multiplies this time of signal travel by the speed of travel to get the distance between the GPS receiver and the satellite. Since these are radio waves, the speed used is the speed of light. One satellite gives a sphere on which the receiver sits. Two satellites give two spheres on which the receiver sits. The intersection of two spheres (and they must intersect) is a circle. Adding a third satellite gives the receiver one of two points at which the sphere will intersect the circle. Using the geoid as the fourth solid, the receiver fixes the point of location. Despite this, there is still some possibility for error if the clock on the receiver has a slight error. A clock error of only one-thousandth of a second causes a position error of almost 200 miles. The solution is to use geometry. If one more satellite is added, then even if the clock in the receiver is off, it is off for all of the satellites by the same amount. The receiver lies on a line from each of the satellites. If all clocks are exact, then the receiver will sit at the intersection of the lines. However, the error in the receiver clock will cause the lines to intersect in different points, resulting in a polygon surrounding the receiver. The receiver can be calculated to be at the center of this polygon.

GPS Capabilities and Accuracy

A GPS receiver must be locked on to the signal of at least three satellites to calculate the latitude and longitude and to track movement. With four or more satellites, the receiver can determine the user's latitude, longitude, and altitude. Once the user's position has been determined, the GPS unit can calculate other information, such as speed, bearing, track, trip distance, distance to destination, sunrise and sunset times, and more. Most GPS receivers are accurate to within 15 meters on average. Newer GPS receivers often come with wide-area augmentation system (WAAS) capability that can improve accuracy to less than three meters on average. No additional equipment or fees are required to take advantage of WAAS. Users can also get better accuracy with differential GPS (DGPS), which corrects GPS signals to within an average of three to five meters. The U.S. Coast Guard operates the most common DGPS correction service. This system consists of a network of towers that receive GPS signals and transmit a corrected signal by beacon transmitters. In order to get the corrected signal, users must have

Once a user's position has been determined, a GPS unit can calculate not just location but speed, bearing, track, trip distance, distance to destination, and sunrise and sunset times, among other things. (Photos.com)

a differential beacon receiver and beacon antenna in addition to their GPS.

Possible sources of error include the following:

- *Ionosphere and Troposphere Delays.* Different layers of the atmosphere have different impacts on the speed of the satellite signal through those layers. Mathematicians have been working on creating better models of these atmospheric layers in order to give smaller errors.
- *Geoid Error.* The receiver uses a mathematical model of the surface of Earth, the geoid. Better mathematical models can improve the accuracy as long as they are relatively easy to use in computation.
- *Signal Multipath.* The GPS signal may be reflected off objects, increasing the travel time of the signal, thereby causing errors. Mathematicians are working on developing models to account for multipath based on the relative location of receiver.
- *Orbital Errors.* Inaccuracies in the satellite's reported location are handled by the control segment, which tries to keep each satellite on track.
- *Number of Satellites Visible.* If only three satellites are visible, the receiver gives a position with a warning that it is likely to be very inaccurate.
- *Satellite Geometry/Shading.* Differences in the relative position of the satellites at any given time may cause errors. Ideal satellite geometry exists when the satellites are located at wide angles relative to each other. Poor geometry results when the satellites are located in a line or in a tight grouping.
- *Intentional Degradation of the Satellite Signal.* Selective Availability (SA) is an intentional degradation of the signal previously imposed by the U.S. Department of Defense. SA was intended to prevent military adversaries from using the highly accurate GPS signals. The government turned off SA in May 2000, which significantly improved the accuracy of civilian GPS receivers.

GPS Signal Transmission

GPS satellites transmit two low-power radio signals, designated "L1" and "L2." Civilian GPS uses the L1 frequency of 1575.42 MHz in the UHF band. A GPS signal contains three different bits of information: a pseudo-random code, ephemeris data, and almanac data. The pseudorandom code is simply an identification code that identifies which satellite is transmitting information. Ephemeris data, which are constantly transmitted by each satellite, contain important information about the status of the satellite (healthy or unhealthy), current date, and time. The almanac data tell the GPS receiver where each GPS satellite should be at any time throughout the day. Each satellite transmits almanac data showing the orbital information for that satellite and for every other satellite in the system.

Further Reading

Cooke, D. *Fun with GPS*. Redlands, CA: ESRI Press, 2005.
Kaplan, Elliot D., and Christopher Hegarty, eds. *Understanding GPS: Principles and Applications*. 2nd ed. Norwood, MA: Artech House, 2005.
Levitan, Ben. *GPS Quick Course: Systems, Technology and Operation*. Fuquay-Varina, NC: Althos, 2007.

DAVID ROYSTER

Green Design

Category: Architecture and Engineering.
Fields of Study: Data Analysis and Probability; Geometry; Measurement.
Summary: Green design requires evaluating the life cycle of a product or material and the cost of that life cycle in energy and other resources.

Green design, also called "environmental" or "sustainable design," is a set of design principles for optimizing environmental impact. This includes reducing pollution, promoting ecological and economical sustainability, using reusable resources, and promoting harmony between people and natural environments. Mathematics plays a significant role in both designing green solutions to a variety of problems and measuring the impact of green solutions. Many colleges offer degree or internship programs in green design, which requires strong science and mathematical skills.

Impact Measures

Ecological design employs a series of metrics for evaluating the degrees of sustainability. A mnemonic used for types of sustainability is "Three Rs": reduce, reuse, and recycle. Reducing waste, pollution, and resource use involves calculations of the impact of production, packaging, transportation, and disposal, as well as renewability of resources. Some design movements, such as Tiny Houses, are predominantly based on the principle of reducing space and resources. Reuse design principles allow objects to be used multiple times, possibly for different purposes. Recycling is the ability to turn objects into materials for making other objects.

The notion of life cycle is central to measuring environmental impact. For example, product life cycles include research and development, main use, and disposal after use. Different stages in the cycle require different types of impact measures. Green design has to address all the stages, from sustainable research practices to possibilities of reuse and recycling at the last stage of the product's life.

There are numerous rubrics and point systems for measuring environmental impacts of industrial, product, or architectural designs. For example, products, activities, or organizations can be measured by their resource intensity, with amount of resources used per unit cost. A toy designer can calculate liters of water spent during manufacture per dollar of the toy's cost. The inverse of resource intensity is resource productivity, measured in quantity or price per unit of resource spent. In this example, resource productivity is the price, in dollars, of toys produced using one liter of water.

Leadership in Energy and Environmental Design (LEED) is an international green building certificate. To give a building or a community its score, LEED combines metrics, such as the carbon footprint, as well as energy and water efficiency. LEED has separate ratings for construction of commercial buildings and homes, interior design, maintenance of existing buildings, and neighborhood development. In each category, the maximum score is 100 points, with certification levels of Platinum (more than 80 points), Gold (60–79 points), Silver (50–59 points), and Certified (40–49 points).

There is a global mathematical problem involved in measuring and reducing environmental impact of design. Namely, there are money and environmental price differences between different design types, and noticeable costs of certification and measurement. The overall sustainability measures have to include all these costs and optimize the total. Because many current economical practices are standardized in nonsustainable manners, the economy of scale makes their use cheaper than the corresponding green designs. This phenomenon is being addressed at the government level by changing price and tax structures to promote sustainable practices.

Green Urban Design

New urbanism is an example of urban design that includes several green principles, including jobs within walkable distances, bike-friendly roads, shared public and housing spaces, diverse communities, and matching local terrain and conditions in landscaping. Geometries of new urbanist designs are concentric and include discernible centers for neighborhoods, such as a historical artifact or a town square, with a transit node tied to this center for optimized logistics. Houses of different types, matching a variety of family and economic situations, are situated within the five-minute walk radius (about one-half kilometer) from this center, and commercial properties surround the houses. The design of roads uses network science to slow down car traffic, minimize travel, and place important administrative, educational, and religious public buildings in traffic network nodes. This relatively compact design, the opposite of urban sprawl, also helps make electricity, water, and gas distribution more efficient, because less energy is spent on delivering these resources and less is lost in transit.

Models from Nature

One of the principles of green design is the use of models found in nature to build products or systems. For example, thermoeconomics models the design of social structures on the laws of thermodynamics. Economical entities are considered on the basis of energy, matter, and information involved in them. Production and use of goods and services are seen as energy and mass exchange, and scarcity has to do with entropy.

The concept of exergy is especially important in industrial design. Exergy is the maximum work theoretically possible as a system reaches energy equilibrium with its surroundings. The second law of thermodynamics says that systems tend to dissipate energy or increase entropy. This loss of exergy is called "anergy." Green designers use both energy and exergy efficiency. Energy efficiency measures how much energy is lost during industrial processes. Exergy efficiency has to do with minimizing anergy, that is, the loss of exergy.

Some social designers consider the total exergy of Earth or even the solar system, working toward designs at these large scales. For example, burning oil or coal produces heat, but these fuels also required inputs of exergy in their making. A mathematical model can approximate the history of the fuels and incorporate their current use, computing energy and exergy efficiency of our actions with regard to Earth, and the sustainability of Earth, over time.

Biomimicry, biomimetics, and bionics are direct uses of design ideas and principles found in nature. For example, engineers studied birds and insects to develop flying devices. More recent examples have to do with efficiency and sustainability. The shape of nautilus shells, mathematically related to the Fibonacci sequence named for mathematician Leonardo Fibonacci, is used to minimize friction in fans, conserving energy. The mechanism of water condensation used by desert beetles can be applied on the human scale. The ways termites keep their mounds warm at night and cool during the day are studied to produce sustainable air conditioning in houses.

Designers and engineers rarely repeat natural designs completely but rather analyze them to find appropriate elements and include elements into the design. There are three directions for such analysis. Designers can incorporate methods of manufacture found in nature, such as the strong material of the mussel's shell. They can mimic mechanical or thermodynamical principles found in nature, for example, the way butterfly wings are colored as the basis of energy-efficient displays. Finally, designers can look at the global organizational principles found in nature, such as modeling a robotic cleaner on insect scavenging behaviors or building artificial intelligence based on the ways brains work.

Further Reading

Andraos, John. *The Algebra of Organic Synthesis: Green Metrics, Design Strategy, Route Selection, and Optimization*. Boca Raton, FL: CRC Press, 2011.

Passino, Kevin. *Biomimicry for Optimization, Control, and Automation*. New York: Springer, 2004.

Vallero, Daniel, and Chris Brasier. *Sustainable Design: The Science of Sustainability and Green Engineering*. Hoboken, NJ: Wiley, 2008.

Maria Droujkova

Helicopters

Category: Travel and Transportation.
Fields of Study: Algebra; Geometry.
Summary: Helicopters apply vertical thrust to overcome their weight.

A helicopter is a type of aircraft that overcomes gravitational force by employing spinning blades to generate vertical thrust. The ideas of vertical flight can be traced back to the Chinese and to Leonardo da Vinci. Thomas Edison studied several different propeller designs and concluded that a feasible helicopter needed a lightweight engine that could produce a large amount of power. Mathematicians such as Theodore Karmen and George de Bothezat also worked on helicopter design in the early twentieth century. In modern helicopters, downward force is supplied by an engine driver rotor. A helicopter has many advantages over a fixed-wing aircraft, such as the ability to take off and land vertically, to hover, and to fly backwards and laterally in the air. As the main rotor spins, it generates a torque that could set the helicopter into a fatal spin. To compensate for this, helicopters have a smaller rotor and blades on their tails.

Flight Controls

A helicopter has four main flight control inputs that enable it to perform various aerial maneuvers: the cyclic control, the collective pitch control, the anti-torque pedals, and the throttle. The cyclic control changes the pitch of the rotor blades cyclically, enabling the helicopter to move in the desired direction. The collective pitch control controls the altitude of the rotorcraft. The anti-torque pedals change the pitch of the tail, altering the amount of thrust.

Mathematically Modeling Helicopter Flight

Helicopters fly by sucking air from above their rotors and forcing it downwards with a thrust equal to (if hovering), greater than (if climbing), or less than (if descending) their weight. The pressures at various points around a helicopter are given by

$$P_0 + \frac{1}{2}\rho v_{out}^2 = P + \frac{1}{2}\rho v_{in}^2 + \frac{1}{2}\rho v_{out}^2$$
$$= P + \Delta P + \frac{1}{2}\rho v_{in}^2$$

here P_0 is the rest pressure of the air far above the rotors, $P + \Delta P$ is the pressure below the rotors, v_{in} is the velocity of the air as it is sucked in, and v_{out} is the velocity of the air as it is forced down.

There are also equations governing the stability and flight of a helicopter. These take into account the inertial velocities in the moving axes system, the Euler rotations defining the orientation of the fuselage axes with respect to Earth, and the aircraft mass. In the early twenty-first century, mathematicians model areas of helicopter flight and performance, such as aerobatic maneuvers that push the limits of the system and that help inform improvements and future designs of new helicopters.

Transverse Flow and Ground Resonance Effects

In forward flight, because the air is being accelerated for a longer period of time as it travels to the rear of the rotor system, air passing through the rear portion of the rotors has a greater downwash angle than the air passing through the forward portion. This pressure difference causes a decrease in the angle of attack, resulting in less lift in the rear of the rotorcraft, increased angle of attack, and more lift in the front. This is called the "transverse flow effect" and it causes easily recognizable vibrations.

When a helicopter is resting on the ground with its rotor spinning, a destructive harmonic vibration called "ground resonance effect" can develop and is caused by a reaction of the rotor blades to the lateral motion of the helicopter. Ground resonance effect develops when the rotor blades move out of phase with each other and cause the rotor disc to become unbalanced.

Further Reading

Ganiev, R. F., and I. G. Pavolov. "The Theory of Ground Resonance of Helicopters." *International Applied Mechanics* 9, no. 4 (1973).
Leishman, J. G. *Principles of Helicopter Aerodynamics*. New York: Cambridge University Press, 2006.
Padfield, Gareth D. *Helicopter Flight Dynamics*. Oxford, England: Blackwell, 1996.
Wagtendonk, W. J. *Principles of Helicopter Flight*. Newcastle, WA: Aviation Supplies & Academics, 2006.

Ashwin Mudigonda

Highways

Category: Architecture and Engineering.
Fields of Study: Algebra; Geometry; Measurement; Number and Operations.
Summary: Highway design requires an adequate model of anticipated traffic and a determination of the grade.

In the early twentieth century, a series of Federal Aid Highway Acts aimed to create a national highway system. Considerations for a highway design include government design specifications and speed limits, the planned route's geographical and geological features, water drainage requirements, land use issues such as environmentally sensitive areas, driver comfort and safety, and maximization of the highway's life span. Planners and engineers also gather data and determine the minimum and maximum expected traffic volumes

Highway Safety

Vehicles traveling along a highway must be able to safely transition between the different gradations and straight sections of a highway. Designers incorporate horizontal and vertical curves to ensure a gradual transition. Designers use mathematical calculations to ensure that the centrifugal forces created by driving along curved surfaces will not adversely affect the vehicle. The calculations involved in designing curves take a variety of data into account, including designed vehicle speeds, geological features, highway and vehicle types, grade, driver sight line obstructions, stopping distance, and connections with other roadways.

based on number of standard axles, vehicle types, expected uses, driver visibility requirements, and the minimum radius of bends and curves. The mathematics used in designing the combination of horizontal and vertical, and straight and curved, sections of a proposed highway results in a design plan that construction crews follow as they build and maintain the highway. Mathematicians also investigate questions related to highways such as mileage, distance, and traffic issues. Mathematician and physicist Louis Roberts served as director of energy and environment at the Transportation System Center in Massachusetts, a division of the U.S. Department of Transportation that researches and develops transportation-related energy conservation practices.

Modeling Highways

Highway designers utilize mathematics to create a three-dimensional layout when planning the horizontal and vertical sections that comprise a highway. The plan view (x and z coordinates) shows the proposed highway's horizontal alignment, which is comprised of straight sections known as "tangents" and the horizontal curves that connect them. The profile view (y axis) shows the proposed highway's vertical alignment, which is comprised of the various slopes known as "grades" at points along the highway. Computer software programs enable modern engineers to create visual models of the plan route and aid in the mathematic calculations involved.

One of the key calculations of highway design and construction is the determination of the necessary grade along the various sections that comprise the highway, defined as the measure of the highway's slope. The grade of a section of highway is calculated using the equation grade = (rise ÷ run) × 100. This equation divides the highway's height increase along that section, known as the "rise," by the horizontal distance a vehicle on a level highway section travels, known as the "run." Designers express distance as stations, whereby one station is 100 feet of highway alignment distance.

The resulting decimal calculation gives the ratio of rise-to-run, which is the grade of that particular section of highway. The decimal grade is then converted to and expressed as a percentage through multiplication by 100. Grade calculations are used to ensure smooth traffic flow along the highway and along the intersections between highways and other roadways as well as to ensure proper water drainage. Designing the proper grade can also help reduce fuel consumption and prevent accidents. During construction, crews move and level the dirt along the right-of-way to create the desired grades.

Vertical Curves

There are two types of vertical curves used in highway design: sag vertical curves and crest vertical curves. The difference between the two is the measurement between the tangent grades at the starting and ending points of the curve, expressed as a percentage. An ending tangent grade that is higher than the beginning tangent grade defines a sag vertical curve, while an ending tangent grade that is lower than the beginning tangent grade defines a crest vertical curve. Thus, a sag vertical curve has a negative value and a crest vertical curve has a positive value. These measurements and calculations combine to create the completed highway design plan.

Further Reading

Clevenson, Larry, Mark F. Schilling, Ann E. Watkins, and William Watkins. "The Average Speed on the Highway." *College Mathematics Journal* 32, no. 3 (2001).

Fwa, T. F. *The Handbook of Highway Engineering.* Oxfordshire, England: Taylor & Francis, 2006.

O'Flaherty, C. A. *Highways: The Location, Design, Construction and Maintenance of Road Pavements.* Oxford, England: Butterworth-Heinemann, 2002.

Schoon, John G. *Geometric Design Projects for Highways: An Introduction.* Reston, VA: ASCE Press, 1999.

Marcella Bush Trevino

HOV Lane Management

Category: Travel and Transportation.
Fields of Study: Algebra; Data Analysis and Probability; Measurement; Problem Solving.
Summary: The decision to designate a traffic lane as a High Occupancy Vehicle lane is based on traffic analysis, computer simulations, and mathematical models showing the effects of implementation.

High occupancy vehicle (HOV) lanes are intended to improve automobile transportation efficiency by reserving certain traffic lanes for vehicles carrying at least two or three people. The idea is to encourage carpooling by allowing cars with multiple occupants to use a dedicated lane and thus reduce the number of cars on the road relative to the number of people traveling. Sometimes traffic lanes are designated HOV only at certain times of the day, or they may be used under special circumstances by buses, hybrid power vehicles, or other single-passenger vehicles. HOV lanes have been tested or used in many countries, including the United States, Canada, Spain, the United Kingdom, Norway, Austria, Indonesia, Australia, and New Zealand. Mathematical modeling, data analysis, and computer simulation are widely used for making decisions regarding when and where to use HOV lanes, for designing their construction and geometric properties, and for evaluating their safety and effectiveness. Many mathematical modelers are using cross-disciplinary concepts and approaches to analyze traffic. For example, engineer Morris Flynn and mathematicians Aslan Kasimov, Jean-Christophe Nave, Rodolfo Rosales, and Benjamin Seibold modeled traffic jams using continuous density and flow functions similar to those used for modeling fluid flow and the propagation of detonation waves. Analogous to the traveling nonlinear wave solutions called "solitons," they christened traffic waves "jamitons."

HOV lanes are typically most useful in regions that have severe traffic congestion and many vehicles carrying only the driver. The opportunity to use less-congested, quickly moving HOV lanes is intended as an incentive to encourage drivers to decide to use carpooling or to carry passengers, with the overall intent of reducing traffic jams and accidents caused by traffic volume and lane changing. Studies of HOV lane usage and effectiveness showed that, as of 2008, 21 U.S. states had HOV lanes for a total of 1,745.14 miles with an average density of 833 vehicles per lane per hour and a total of over 276 million miles of vehicle travel. Exclusive HOV lanes were most common (993.27 miles) and carried the highest density of traffic (an average of 906 vehicles per hour), followed by normal lanes designated HOV in certain periods (545.82 miles, 790 vehicles per hour) and shoulder or parking lanes designated HOV in certain periods (206.6 miles, 596 vehicles per hour).

Experience with HOV lanes is mixed, although it should be noted that this is a relatively new method of organizing transportation and that local variation in conditions and implementation could explain why some projects were more successful than others. An example of a successful HOV implementation was that introduced in 1998 near Leeds, United Kingdom (the first HOV lanes in the United Kingdom). Prior to HOV lane implementation, 30% of the cars had two or more occupants, and a journey that should take three minutes if traffic were moving freely regularly took more than 10 minutes. After implementation of the HOV lanes, traffic was reduced 10% to 20%, journeys were quicker for both HOV and non-HOV traffic, lane violations were low, casualties were reduced 30%, and noise reduction was noticeable—although little change was noted in air quality. In the United States, an HOV lane scheme near Washington, D.C., for vehicles carrying four or more occupants, proved successful, with the HOV lanes operating at twice the speed of travel of the regular lanes. However, a study of HOV lanes in San Francisco, California, found that they actually increased congestion. HOV lanes have also been criticized on grounds of safety, because of the differing speeds of traffic in adjacent lanes, and as a violation of the right of motorists to freely use highways paid for with their tax dollars.

Mathematicians continue to investigate issues for HOV lane design, implementation, and management. Analyses using concepts from fields such as geometry, graph theory, and statistics help designers optimize features like lane setbacks, entrance and egress paths, gates and signals, and shoulder widths. Speed contour plots can be used to visualize recurrent blockages, while probability models and scatterplots can be used to quantify and display spatial distribution of accidents as functions of one or more variables. Other mathematicians seek to simplify existing multiparameter models, which may rely on unobservable quantities, using smaller sets of physical and measurable variables in order to study the impact of design features and traffic behavior. Yet others have used logit-type models to investigate economic concerns, like converting HOV lanes to high occupancy toll (HOT) lanes.

Further Reading

Kwon, Jaimyoung, and Pravin Varaiya. "Effectiveness of High Occupancy Vehicle (HOV) Lanes in the San Francisco Bay Area." *Transportation Research Part C: Emerging Technologies* 16, no. 1 (February 2008).

http://paleale.eecs.berkeley.edu/~varaiya/papers_ps.dir/HOV.pdf.

Menendez, Monica. *An Analysis of HOV Lanes: Their Impact on Traffic*. Saarbrücken, Germany: VDM Verlag, 2008.

U.S. Department of Transportation, Federal Highway Administration. "High Occupancy Vehicle (HOV) Lanes by State." http://www.fhwa.dot.gov/policyinformation/tables/03.cfm.

Sarah Boslaugh

Internet

Category: Communication and Computers.
Fields of Study: Algebra; Number and Operations; Problem Solving.
Summary: Many properties and problems of the Internet are studied and modeled using mathematics.

The Internet is a worldwide computer network connecting other computer networks in government, business, academia, and other public and private sources. Communications are facilitated by the Internet Protocol Suite (TCP/IP), originally proposed by Vinton Cerf and Robert Kahn in 1974. The Internet is used for implementing various applications including electronic mail, pioneered in the late 1960s, and the World Wide Web (WWW) of linkable documents. The idea of networks connecting information nodes appeared in futuristic scientific writings and science fiction beginning in the early twentieth century.

The work of mathematicians, computer scientists, cyberneticists, and many other scientists contributed to the emergence of the Internet and the World Wide Web by the end of the twentieth century. Researchers and teachers in nearly every discipline use the Internet to further their work, and many study the properties of the Internet itself using mathematics. One problem explored by mathematicians and computer scientists is mapping the Internet, often undertaken to understand the nature of connections and to reduce stress on routers. The field of hyperbolic geometry has proven to be highly useful in creating such maps, especially with regard to assessing global stability and developing efficient routing methods. Mathematicians also consider the theoretical and computational challenges posed by the massive graphs that result from Internet mapping, which test the limits of even the largest and fastest computers. Others examine society's increasing dependence on the Internet for a range of critical everyday tasks (like banking and medical recordkeeping) along with the risks and vulnerabilities (like identity theft) that this reliance may create.

Codevelopment of Mathematical Sciences and the Internet

Mathematicians including John Von Neumann, Alan Turing, and Norbert Wiener contributed to the development of both the hardware and the software necessary to implement computer networks and the Internet. The precursors of the Internet were networks such as the telegraph, telephone, radio, and television. Even early electronic computers had systems for data input, computation, and output. In the late 1960s, individual computer "nodes" were connected to one another, building on the technology for connecting subsystems within the same computer. These early stages of building computer networks promoted the development of the mathematics-rich fields of cybernetics, informatics, and artificial intelligence.

Mainframe computers enabled countless historical achievements and facilitated research and problem-solving in mathematical fields such as cryptography, simulation, and genetics. In the late 1970s and early 1980s, the introduction of the first personal computers changed the face of computing by creating applications and giving access to new groups of users. In the 1980s, the National Science Foundation (NSF) funded five supercomputer centers connected by NSFNET, which built on Computer Science Net (CSNET) and the Department of Defense's Advanced Research Projects Agency Network (ARPANET). Demand during the first year was so great that the system had to be upgraded almost immediately, and uses for the new network continued to expand, as did the mathematics research needed to meet user demands for functionality. At the same time, national computer networks such as ARPANET and NSFNet, the Japanese JUNET, and the European CERN remained isolated from one another. The big challenge at the time was to make these separate networks compatible and interoperable.

Adoption of several dozen international protocols, such as the TCP/IP for the Internet, facilitated interlinking. In the early 1990s, the idea of common protocols enabled the system of file hosting, accessible by anyone at all times and called the World Wide Web. The explosive evolution of the Internet and the Web in the next decade is well documented. In the United States, efforts were aided by several pieces of legislation. For example, the High Performance Computing Act (HPCA) of 1991 reset priorities for computing research and education. President Bill Clinton stated that he believed such legislation enabled collaborations "critical for assuring American prosperity, national and economic security, and international competitiveness in the twenty-first century." Computer scientists Eric Bina and Marc Andreessen developed the first widely used graphical browser, Mosaic, released in 1993 and funded by a program associated with the HPCA. Tim Berners-Lee, the creator of several WWW protocols, was knighted in 2004 by Queen Elizabeth II "for the invention of the World Wide Web."

Mathematical Problems

One mathematical problem that had to be solved in order to build computer networks was packet switching, which is grouping data of all types into blocks known as "packets" of size that are appropriate for network transmission. Network nodes or routers have algorithms that decide how to queue, buffer, and deliver individual packets as a function of network traffic patterns. This is a different, mathematically more complex model from circuit switching, which was used in older telephone networks to transmit information bits at a constant rate. Computer scientists Paul Baran, Donald Davies, and Leonard Kleinrock pioneered packet switching networks. Baran's work was shaped in part by Cold War concerns about maintaining communications in the face of nuclear attack. Donald Davies worked with Alan Turing at the National Physical Laboratory and is reputed to have found mistakes in Turing's groundbreaking paper "On Computable Numbers." Kleinrock, a recipient of the U.S. National Medal of Science, said of his work, "Basically, what I did for my Ph.D. research . . . was to establish a mathematical theory of packet networks."

In the late 1960s, mainframe computers had message systems among their different users, who all had to be online at the same time to communicate. In the early 1970s, the message system software was modified to include new computer networks. The ability to deliver messages to offline users, make different systems compatible, and uniquely identify users were significant research problems. The compatibility issue, still important in the twenty-first century, was resolved in part by creating software and hardware gateways that connect different systems. BITNET was cofounded by Ira Fuchs and Greydon Freeman primarily for research and academic communities, while FidoNet was implemented for personal computers and bulletin board systems by Thomas Jennings. Unique identification of users is a complex mathematical problem, since for any string length there is a finite number of possible letter and symbol permutations.

Similar concepts apply to the study and selection of secure electronic passwords. A system developed in the early 1970s assigned registration codes to domains and then to users within domains in the form "user@domain." This method and the use of "@" are credited to Raymond Tomlinson. At the start of the twenty-first century, the mathematical structure of domain names is a type of tree, with multiple hierarchical levels. Minimally, there are two levels. Each domain name ends with the top-level domain including generic ones, such as ".com" and ".edu," and country code ones, such as ".us" or ".uk," with a period on the left. To the left of that period comes the second level domain name; for example, "wikipedia.org" or "google.com."

If there are more domain levels, they appear on the left of the second-level domain and are separated by periods as well; for example, "simple.wikipedia.org" or "groups.google.com." There is no limit to the number of domain levels. This syntax and structure was first published in the 1980s in connection with the Advanced Research Projects Agency Network (ARPANET). IP addresses are the numerical representations of individual computers, mapped to domain names. They consist of four bytes of information displayed as numbers. Each byte has eight bits and can be any integer from 0 to 255. With the exponential growth in Internet users, assigning unique identities to users, domains, and computers continues to be a challenging problem, especially since many users have multiple e-mail and IP addresses. For computer users, off-line message delivery is achieved by storing messages on digital servers until the recipient accesses them.

E-mail programs typically employ the Internet Message Access Protocol (IMAP), developed by Mark Crispin, or the older Post Office Protocol (POP) to

A mathematical problem that had to be solved in order to build computer networks was packet switching—grouping data of all types into blocks known as "packets" that are sized appropriately for network transmission. (Photos.com)

retrieve mail. Simple Mail Transfer Protocol (SMTP) is also used for sending and receiving functions. Mathematical algorithms enable the queuing, encryption, authentication, and filtering of e-mail, and mathematicians continue to contribute new developments and improvements. Many agencies are responsible for making assignments and tracking Internet protocols. The Internet Assigned Numbers Authority was headed for nearly 30 years by computer scientist Jonathan Postel, who codeveloped and documented many of the key Internet standards, including SMTP and Domain Name System (DNS) servers.

The Growth of Networks

Other mathematical problems of Internet development sprang from the incredibly fast growth of networks. To compare the rate of growth of different networks, researchers use metrics such as time per number of users. They have determined, for example, that it took only five years for the Internet to reach 50 million users, versus 13 years for television and 38 years for radio. As the number of users and domains grew, search algorithms became a prominent field in computer science and mathematics, with several major developments such as clustering and relevance rankings. There are many search engines, many of which initially used the content of Web pages to rank results. Google's PageRank method was among the first search protocols to use sophisticated mathematical modeling, including directed graphs and stochastic matrices, to explore links between pages hierarchically. The Page Rank algorithm is named for Google cofounder and computer scientist Lawrence Page.

In 2009, Google research scientist Kevin McCurty noted that successful search engines continually improve by employing mathematical methods that quickly find relevant material and eliminate irrelevant factors that can skew results. Along with better ranking schemes, Internet speed is critical in effective searching and content delivery. The original packet switching and data routing problems have become even more complex as the Internet has grown. Mathematicians and computer scientists model Internet traffic flow using many mathematical and statistical techniques, taking into consideration many variables, including the type of content being exchanged. Photos, videos, music, text, e-mail, and online gaming all require different resources. Based on these models, algorithms to optimally route traffic can be designed and implemented, reducing congestion and slowdowns. For example, the traffic load on a given Website's computers can be reduced by storing some content at other servers that provide more optimal access patterns, a process known as "network caching."

Some twenty-first century models are starting to use concepts from disciplines like economics, such as equilibrium theory. One example is called "congestion-dependent pricing," which would route packets depending on users' willingness to pay more for privileged Internet access during periods of congestion. Given the number of packets in even a small text file, this is a mathematically complex problem that still requires a great deal of research.

A separate set of science problems has to do with hardware and the various means of connecting to the Internet. As of 2010, it is possible to connect to the Internet through both land-line and cell phones, radio, satellites, dedicated fiber-optic lines, and television cables. While similar in many ways, each has a unique set of issues related to speed, security, data transmission, compatibility, and bandwidth, especially when considering that people are connecting to the Internet with many devices other than personal computers. Mathematicians, computer scientists, and others work on both the hardware and the software solutions.

Network Science

Network science predates the Internet, having its root in graph theory. It is interdisciplinary and includes mathematics, engineering, computer science, biological sciences, sociology, and other disciplines interested in studying various types of networks. It flourished with easy availability of empirical data from computer and social networks made possible by the Internet and the high demand for applications in all aspects related to the Internet. Concepts and methods from graph theory, such as centrality, betweenness, and closeness are used to quantify and describe networks. Centrality is a measure of the importance of a node within a network. Betweenness measures the quality of paths through the node, such as the number of shortest paths between pairs of other nodes. Closeness is the topological measure similar to distance, usually defined as the average number of nodes in the shortest path between a given node and all other nodes in a network that connect to it.

Maps of networks help mathematicians and others analyze vulnerabilities, such as critical nodes that lie between many other nodes and whose loss would sever connectivity, and deprecated connections, where use of outmoded software or features affects speed or leaves the users open to attack. In addition to graph theory, hyperbolic geometry adds to Internet mapping by considering geometric coordinates of nodes in space, not

Erdös–Rényi Graphs

One mathematical discovery of network science is that large-scale networks like the Internet are structured in ways that do not appear to be random, though some researchers initially thought they would produce Erdös–Rényi graphs, which are random graph models having bell-shaped degree distributions. They are named for mathematicians Paul Erdös and Alfréd Rényi. Instead, large social networks have degree distributions with no peaks and heavy tails, proportional to a power function. This means that most nodes have very few connections, and only a few nodes, called hubs, have many connections.

For example, the majority of Wikipedia editors have edited only one or two articles, and the majority of Web pages have one or no links leading to them. In 2004, physicist Mark Newman and his colleagues studied scientific coauthorship networks using these models. Mathematician Paul Erdös, acknowledged by many as one of the founders of graph theory, was a highly prolific collaborator—a node of very high degree in the network of published mathematicians, who often compute their personal Erdös number to describe their closeness to Erdös.

simply the map of connections. The added information can then be used to quantify the issue of closeness from a geometric point of view. In graph theory, each node of a network has a degree, which is the number of other nodes connected to it. Degree distribution is a statistical measure showing the probability distribution of various node degrees over the network. Statistical sampling strategies are often used in network research, since the problems and networks examined are typically far too vast for complete data collection.

Economics and the Internet

In the 1990s, many people believed that the Internet would bring about fundamental changes in the landscape of the business world. Starting in the mid-1990s, venture capitalists were investing heavily in new Internet businesses, sometimes called "dot-coms." During this time, many Internet companies operated at annual losses, expanding in anticipation of future revenues. This worked for relatively few companies, such as Amazon and Google. In 2001, this "dot-com bubble" burst, with many Internet-related businesses declaring bankruptcy.

The promises of the Internet that survived the dot-com bubble became clearer toward the end of the first decade of the twenty-first century. For example, researchers found that in many cases, product popularity obeys a frequency distribution law similar to the degree distribution of network nodes. The majority of customers use a few most popular products, with the majority of products liked by small minorities of customers.

In the early 2000s, several companies realized large profits by reaching these so-called long tails (named after the characteristic shape of the distribution curve) of niche customers and redefined their industries. Apple changed the music industry by selling individual tracks online; Netflix had a similar effect on movie rentals. Mathematical algorithms for determining customer preferences and making recommendations were driven in large part by Internet commerce. Recommender systems use complex relevance metrics, evaluating content such as texts or video based on statistics of past behavior of all users within the system.

These systems use explicit data, such as rank preferences given by users, as well as implicit data, such as actions other similar users have done before. Over time, these systems accumulate large amounts of data and increase the accuracy of their recommendations. Mathematics involved in creation of these algorithms includes statistical analysis and linear algebra for working with matrices defining closeness of users. Illustrating how lucrative good algorithms are from the business perspective, in 2009 the Netflix Prize awarded $1 million to the developers of an improved filtering algorithm for recommending movies.

Further Reading

Abbate, Janet. *Inventing the Internet*. Cambridge, MA: MIT Press, 2000.

Boguñá, Marián, et al. "Sustaining the Internet with Hyperbolic Mapping." *Nature Communications* 1, no. 1 (September 2010).

Churchhouse, R. F. *Codes and Ciphers: Julius Caesar, the Enigma, and the Internet*. Cambridge, England: Cambridge University Press, 2001.

Dietrich, Brenda, Rakesh Vohra, and Patricia Brick. *Mathematics of the Internet: E-Auction and Markets*. New York: Springer, 2010.

Langville, Amy, and Carl Meyer. *Google's PageRank and Beyond: The Science of Search Engine Rankings*. Princeton, NJ: Princeton University Press, 2006.

Srikant, Rayadurgam. *The Mathematics of Internet Congestion Control*. Basel, Switzerland: Birkhauser, 2003.

MARIA DROUJKOVA

Landscape Design

Category: Architecture and Engineering.
Fields of Study: Algebra; Measurement; Problem Solving.
Summary: Landscape design is an application of geometry, shaping an outdoor environment to something pleasing.

Landscape design is the combination of gardening and architecture for making outdoors environments more aesthetically pleasing, ergonomic, and useful.

It is a synthetic occupation, requiring the knowledge and skills of horticulturists, engineers, architects, and visual artists.

Mathematical calculations underlie many aspects of landscape design, such as how many plants are needed to fill a bed or landscape or how to build a landscaped terrace that will resist erosion. Landscape architects often include design elements based on symmetry and other geometric features of areas, surfaces, and three-dimensional elements. More advanced mathematical forms, such as fractals or labyrinths, are incorporated in some landscapes, like crop circles. Peter Schaar was an applied mathematician for many years before turning to a career as a landscape designer. He noted that the notion of an elegant solution is common to both mathematics and garden design.

Design Elements and Principles

Landscape design, like other forms of design and decorating, uses design principles and elements that are mathematical in their nature. The Western traditions of landscape design typically use lists of elements, including the following:

- Line
- Shape
- Size
- Texture
- Color

Every element is expressed through natural or architectural media, including plants, stones, and ground shapes. Straight or curved lines and shapes are created using hedges, paths, flower borders, and shapes of bushes and trees. Sizes of landscape elements, including stones, plants, and built structures, can match or contrast. Textures and color can be natural, such as foliage, water, grass, and stone, or modified by people, such as cut bushes, polished stones, and painted structures.

Likewise, the artistic principles, such as repetition, balance, and focal points, are achieved with the combination of human-made and natural elements. For example, traditional landscaping focal points include sculptures, fountains, and flower beds.

Sacred Traditions and the Development of Mathematics

Building, gardening, and designing landscapes were connected to spiritual practices by many cultures around the world. The resulting complexity of habitats often elevated mathematical and scientific knowledge, as well as the arts within the cultures practicing these traditions.

For example, feng shui is the Chinese design tradition connected with the development of astronomy and precise measurement instruments, such as magnetic compasses and astrolabes. Mathematical ideas involved in feng shui symbols include binary numbers, powers, and combinatorics.

Some mid-African cultures use fractal structures in village design, where the shape of the village is repeated in shapes of house clusters, then houses, then rooms within houses. The shape is connected to the beliefs of the people and reflected in the lore while at the same time being practical for the needs of the village.

Ancient Egyptians used the concept of *gnomon*, which is a specially constructed geometric shape corre-

Photos.com

Labyrinths and Mazes

A labyrinth is an elaborate landscape structure consisting of live or stone hedges or mosaic ground patterns, with a winding single path leading to its center. Unlike mazes, which have many possible paths and serve as spatial puzzles, labyrinths are easy to navigate. Labyrinths are used as pleasing places for conversation or meditation. Mathematically similar patterns for labyrinths appear in archaeological finds on all continents. Solutions of mazes and constructions of labyrinths have to do with topology, graph theory, and knot theory in mathematics.

sponding to a regular polygon, in their area and architecture calculations. When a gnomon is added, the ratio of polygon sides is maintained. Osiris was associated with this idea of the constant ratio, in the myth as the God of Sun, growth, and constant change, and was often drawn on a square throne expanded with the L-shaped gnomon. These geometric traditions were inherited by the Greeks, formalized as Euclid's geometry, and entered the Western knowledge base.

Budgets and Rates

Landscaping expenses include the price of material and labor for construction and maintenance. It is estimated that in the United States, a house with its landscape design rated "excellent" by experts can sell for 5% to 10% more than the same house with its design rated "good." Therefore, it may make financial sense to spend money landscaping the property. These calculations are performed by developers and real estate agents when deciding landscaping budgets.

Further Reading

Agnew, Michael, Nancy Agnew, Nick Christians, and Ann VanDerZanden. *Mathematics for the Green Industry: Essential Calculations for Horticulture and Landscape Professionals*. Hoboken, NJ: Wiley, 2008.

Ferrater, Borja, and Carlos Ferrater. *Synchronizing Geometry: Landscape, Architecture & Construction*. New York: Actar, 2006.

Winn, Becky. "The Mathematician's Garden." *Dhome*, February 2006. http://www.dmagazine.com/Home/2006/02/16/The_Mathematicians_Garden.aspx.

Maria Droujkova

Levers

Category: Architecture and Engineering.
Fields of Study: Algebra; Geometry.
Summary: Levers negotiate forces in ways useful in engineering.

Levers are rigid beams that pivot around a point called the "fulcrum" to mediate three forces: an applied effort, a load to be moved, and the fulcrum's reaction. Depending on how the load, effort, and fulcrum are placed along the beam, either force or travel distance can be increased and the other decreased in proportion. There are three classes of lever, distinguished by the placement of the effort, load, or fulcrum. Levers of the first class have the fulcrum between the effort and load, like a see-saw, for changing direction of force and travel distance and increasing or decreasing either of them. The second class has the load in the middle, like a wheelbarrow for increasing force. The third class has the effort in the middle, like a pair of tongs for increasing travel distance. As these examples illustrate, levers are everywhere in the mechanical world and have been for the entirety of civilization.

Levers also occur in animals: the bones in limbs function as rigid rods and fulcrums, with muscles pulling hard close to a joint (the fulcrum) to move the extremity through greater distances than the contracting muscle can cover but exerting a force weaker than the muscle exerts on the bone. A train of three levers—the hammer, stirrup and anvil bones—magnify tiny acoustic displacements as they transmit sound from the eardrum to the cochlea.

Early Study

Our present formulation of levers derives from the *Equilibrium of Planes* of Archimedes, who determined that "Magnitudes are in equilibrium at distances reciprocally proportional to their weights." Using levers, Archimedes investigated the volumes of spheres and cones. Archimedes imagined the cone or sphere divided into thin slices: if a slice is hung on one side of a lever, what cylinder slice must be hung at what position to maintain equilibrium? By working through the entire volume of the cone or sphere, Archimedes constructed a cylinder of equal volume, thus giving the sphere's and the cone's volume. Levers also appear in Galileo's 1638 book of mechanics, *Two New Sciences*. Whereas Archimedes had abstracted the lever as a perfectly rigid line, Galileo considered it as a three-dimensional, flexible object, leading to the first theory of beams. Combinations of levers, constrained in various ways, became a research topic during the Industrial Revolution. "Linkages," as these devices are called, were important for converting the rotation of steam engines into linear motion. Researchers in the nineteenth century took a mathematical approach to the problem. Among the best-known linkages is the Peaucellier cell, invented in 1864. The Peaucellier cell also plays theoretical roles in computer science.

Applications

Levers feature in mobiles and, notably, in the sculptures of Alexander Calder, who often places the fulcrum slightly above the beam that assists in balancing. The raised fulcrum has long featured in balances for weighing; the pivot point is above the lever's center of gravity so that, when the pans pull with equal torque, torque from the displaced beam's own weight will pull it level. Not all balances rely on this feature. Chinese pharmaceutical balances, for example, require the operator to look for nonrotation rather than perfect leveling.

More generally, nonmechanical levers exploit length to multiply distance. Optical levers rely on a mirror doubling an angle and a long travel distance for the light ray to register a large displacement. Social, financial, intellectual, and political resources can be metaphorically "leveraged" by using them to achieve outcomes larger than the resource itself, though the metaphor generally neglects to acknowledge the loss required for a mechanical lever to provide any gain.

Further Reading

Heath, T. L., ed. *The Works of Archimedes*. New York: Dover, 1953.

Moon, Francis C. *The Machines of Leonardo da Vinci and Franz Reuleaux*. Dordrecht, Netherlands: Springer, 2007.

Reynolds, Laura. "How to Build Levers and Pulleys." eHow. http://www.ehow.com/how_4466340_build-levers-pulleys.html.

Alistair Kwan

Light Bulbs

Category: Architecture and Engineering.
Fields of Study: Algebra; Connections.
Summary: Light bulbs are ideally designed for great luminous efficacy, emitting more light than heat.

Light bulbs are common sources of electric light. The light bulb's evolution is not entirely certain. Historians cite more than 20 contributors, dating back to roughly 1800, who made discoveries prior to inventor Thomas Edison's 1879 patent for an incandescent bulb. Some attribute Edison's success to the fact that he also invented an entire electricity distribution system.

Traditionally, light bulbs work on the principle of incandescence. The filament inside an incandescent bulb resists the flow of electrons supplied by an electrical source, causing the filament to heat up and emit radiation. Approximately 90% of the power consumed by an incandescent light bulb is, in fact, emitted as heat rather than as visible light. The wavelength of the emitted radiation determines the color of the light. In common household incandescent bulbs, the emitted radiation is primarily in the infrared region of the spectrum, which humans cannot see, along with the visible red, orange, and yellow wavelengths nearest the infrared. This characteristic gives the bulb its characteristic yellowish color.

Compact fluorescent light bulbs, which are intended to replace incandescent bulbs, operate on a different principle. Electricity excites mercury vapor to produce light, but little heat. The emitted spectrum of traditional fluorescents is much closer to the blue end of the visible spectrum, though there are now a variety of models that closely mimic natural light. In addition to quantifying the emitted radiation spectrum, mathematics is used to calculate other important features of light bulbs, such as electrical rating and efficiency.

Rating

Incandescent bulbs are normally rated according to their electrical power. Common household sizes in the United States range from 15 watts, often found in refrigerators and other appliances, to 150-watt bulbs used for reading or to light large areas. As the bulb is purely resistive (its inductance and capacitance are insignificant), the electrical power can be computed as $P = V \times I$, or $P = I^2 \times R$, where P is the electric power in units of watts, V is the potential difference in volts, R is the resistance of the filament in ohms, and I is the current in amperes or "amps," named after André-Marie Ampère, a French mathematician and physicist considered the "father of electrodynamics." Household voltage in the United States is usually 120 volts, so higher wattage bulbs require more current to operate, which makes them more costly to use. Because compact fluorescents operate on a different principle than resistance, they typically draw less current to produce the same perceived intensity of light.

Luminous Efficacy

Another metric used to distinguish light bulbs is luminous efficacy, defined as

$$\text{LES} = \frac{F}{P}$$

where F, the flux in lumens, is the total useful amount of visible radiant light, and P is the power. A weighted luminosity function adjusts for the human eye's response to different wavelengths of light when flux is calculated. If total electric power consumed by a bulb is used in this computation, it is referred to as "luminous efficacy of a source" (LES). LES is a good indicator of source's ability to provide visible light from a given amount of electricity. For example, a 40-watt incandescent bulb has an LES of roughly 12.6 lm/W, and a flux comparable to a 9- to 13-watt compact fluorescent. A 100-watt bulb has a flux comparable to 17.5 lm/W, versus a 23- to 30-watt compact fluorescent.

Humor

Light bulbs are also a source of humor, with hundreds of light bulb jokes of the general form, "How many (fill in the blank) does it take to screw in a light bulb?" Many of these jokes are intended to satirically poke fun at the subjects; mathematicians are no exception. For example, "How many mathematicians does it take to screw in a light bulb?" The answer is "None. A mathematician can't screw in a light bulb, but he can easily prove the work can be done."

Further Reading

Collier, James L. *Electricity and the Light Bulb.* Tarrytown, NY: Marshall Cavendish Benchmark, 2006.

Kaufman, John. *IES Lighting Handbook 1981 Reference Volume.* New York: Illuminating Engineering Society of North America, 1981.

Ashwin Mudigonda

Mapping Coastlines

Category: Space, Time, and Distance.
Fields of Study: Algebra; Geometry; Measurement; Number and Operations; Representations.
Summary: Fractals can be used to help map coastlines.

A map is an infographic representing an area. Maps use symbols to represent objects or scale renderings of spatial features. The science of mapmaking is called "cartography." The mapping of coastlines is important for navigation and for determining the boundaries of territorial waters, which are measured as fixed distances from coastlines. Coastline cartography presents special mathematics because of connections with several actively developing branches of mathematics, including fractal theory.

Traditional Mapmaking Mathematics and Analytical Cartography

Several mathematical features of maps have been used for centuries. Orientation is the correspondence of the map's coordinate system with directions of the terrain. When three-dimensional objects are depicted in two-dimensional media in the process called "projection," such as maps of Earth on paper, some areas are necessarily distorted. Ratios are used to map objects to scale, including the systematic changes in the ratios in different parts of two-dimensional maps using projections.

With the increasing use of computers in cartography, several new areas of modeling and computation expertise have appeared over the last few decades. These new, mathematics-rich cartography areas include computer-based geographic information systems, interpolation, and photogrammetry. Collectively, these areas of expertise are called "analytic cartography."

Types of data in analytic cartography include numerical data, such as elevation values, images or photographs, and attribute data, like tags identifying features near particular coordinates. All data are dynamically linked and manipulated in a geographic information system; for example, a projection map can be generated from a series of aerial photos, rotated and zoomed. In contrast, paper maps do not allow dynamic data connection and are static, which limits the possibility for mathematical modeling and experimentation with variables. Geographic information systems may also include remote sensing data; for example, displaying changes in coastlines in real time as tides change.

Analytic Cartography and Coastline Changes

Because coastlines change a lot compared to other map features, from tides and floods, analytic cartography that allows for rapid analysis of real-time data is especially valuable in mapping coastlines. Using data from previous events and mathematical models within geographic information systems, cartographers can simulate floods, tsunamis, or effects of rising water levels from global warming on existing coastlines. The same software can be used to predict effects of terrain modification projects over time.

Modeling coastline changes is more complex than simply mapping higher or lower water levels onto the existing coast elevation data. The models also have to take into account erosion, deposits of matter by rivers and rainfall runoff, changes in river basins, and other systemic factors.

Fractal Dimension and the Coastline Paradox

A fractal is a self-similar structure that looks the same at all zoom levels. Coastlines, while not perfectly fractal (not having infinite number of levels), exhibit enough fractal features to make some mathematics of fractals applicable. The famous 1967 paper by Benoît Mandelbrot, "How Long Is the Coast of Britain? Statistical Self-Similarity and Fractional Dimension" started this line of thought, though the term "fractal" appeared later.

An important feature of a coast is its fractal dimension (a measure of how long the coast is compared to the area it occupies). Because the area has two coordinate dimensions and the length has one, theoretically, a curve filling a unit of area can have infinite length. Fractal dimension is a way to compare different coasts, from straight coastlines that have the fractal dimension of 1 to increasingly complex, space-filling coastlines that have higher fractal dimensions between 1 and 2. In Mandelbrot's paper, the relatively smooth coast of South Africa has the fractal dimension of 1.02 and the highly irregular (long for its area) coast of Britain has the fractal dimension of 1.25.

The length of the coast and its fractal dimension depend on the units of measure. Because smaller units allow the cartographer to capture more detail of the coastline, measuring with smaller units produces higher total length. This is definitely not true about measuring straight lines, and thus it is called the "coastline paradox."

Randomness and Pattern

Perfectly self-similar fractals created by mathematical models have limited applicability to coastline mapping because real coasts are irregular. Therefore, some mathematical models include the element of randomness in creation of factors and use statistical methods to compute fractal dimensions. For example, one method for generating random fractals is called "random midpoint displacement," produced by using the following cyclic algorithm repeatedly:

- *Step 1*: Start with a straight line.
- *Step 2*: Displace the midpoint randomly, perpendicular to itself, by the distance within the given ratio to its length.
- *Step 3*: Apply Steps 1 and 2 to the segments resulting from the previous steps.

A similar method can be applied to generating elevation of areas. In this case, the algorithm starts with a rectangle, displaces its midpoint, and then is applied to the four rectangles formed by the lines parallel to the original rectangle's sides and crossing at the midpoint.

Because these methods are computationally intensive, as the number of computations at each step grows exponentially with the number of cycles, their development coincides with increases in computing power. In addition to mathematical modeling of existing coasts, these methods are used to generate fictional terrain for computer games, virtual worlds, and digital artworks.

Coast-Mapping Satellites

Several government and private projects connect real-time satellite data to specialized coastline geographic information systems. This connection provides either real-time or within-minutes data for ship navigation charts, environmental hazards (like oil spills in harbors), and natural disaster data (like tracking tsunamis).

Satellite mapping has to use methods beyond optical imagery because data have to come during the night as well as in cloudy conditions. Coast-mapping satellites use radar sensors that do not depend on light. These sensors measure changes in reflected radar pulses. Rougher surfaces reflect differently from water, allowing for relatively precise mapping of the coastline.

Further Reading

Monmonier, Mark. *Coast Lines: How Mapmakers Frame the World and Chart Environmental Change.* Chicago: University of Chicago Press, 2008.

Seppala-Holtzman, D. N. "International Waters: An Anatomy of an Analysis." *Math Horizons* 14 (September 2006).

Turcotte, Donald. *Fractals and Chaos in Geology and Geophysics.* 2nd ed. Cambridge, England: Cambridge University Press, 2010.

Maria Droujkova

Mattresses

Category: Architecture and Engineering.
Fields of Study: Algebra; Geometry; Measurement.
Summary: Modern mattresses are superior to older designs because of the geometry and pressure distribution of the coil springs that define them. Mattresses last longer when rotated through four configurations.

The modern mattress is a cushion for sleeping and sits on top of a box spring that provides support and reduces wear and tear. While straw, coconut fiber, horsehair, feathers, pea shucks, and water have all been used to stuff mattresses in the past, in the twenty-first century most are filled with artificial fibers or foam rubber and derive much of their resilience and support from coil springs, which are either connected by interconnecting wires or encased in fabric. The first innerspring mattress is attributed to Heinrich Westphal in 1871, and its popularity may be because of hygiene and comfort considerations. The geometry of the coils impacts the durability and firmness. Manufacturers use calculations like the average load limit of a floor and the volume and weight of a waterbed. NASA attributes the 1960s invention of soft memory foam with high-energy absorption properties to aeronautical engineer Charles Yost, who was working under a NASA contract. Many studies employ statistics, such as those involving quality of sleep, amount of snoring, and the impact of sleeping positions. Mathematical techniques and models of mattresses have also been useful in studying factors such as pressure distribution, deformation, combustible behavior, and mattress flipping.

The gauge of the coils is one of the factors that impacts the mattress's firmness and, therefore, its support and durability. Counterintuitively for the layman, lower gauges mean larger cross-sectional diameters—the number of passes through the drawing dies that are required to create a wire of a given thickness. Lower gauge means fewer passes, meaning thicker wire. Bonnell coils, named after the inventor, may have been adapted from the coils used for buggy seats in the nineteenth century. The configuration of adjacent hourglass coils connected by helical wire, called "helicals," increases the spring's resistance proportionally to the load. Cylinder pocket springs systems are individual cylinder pieces held together by clips. Continuous coils are rows formed from a single piece of wire. However, while the head-to-toe rows of continuous coils offer good support while still responding to shifts of position and weight, the movement of the coils in response to those shifts is noisier and more noticeable than in other mattresses.

Wear and tear on a mattress is disproportionate because of the fact that the sleeper and the mattress are not the same shape, and thus some coils will bear more load than others. Compressing and decompressing gradually weakens a mattress; it may show noticeable changes after a few years. Use of a firm box spring helps prevent the sagging that would set in quickly, and

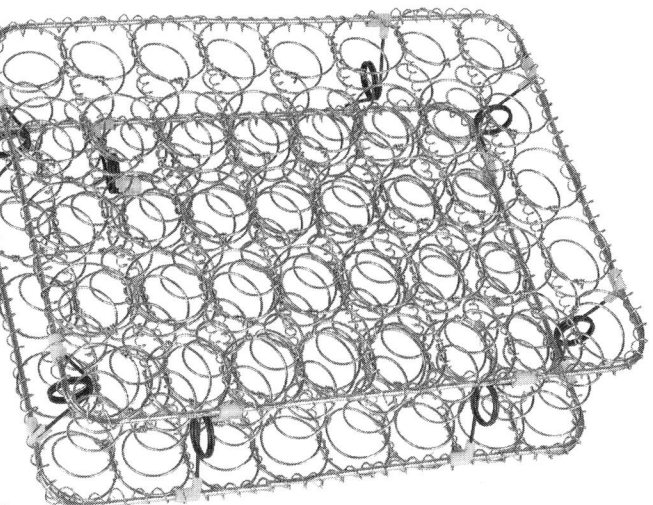

The inner workings of a mattress include coil springs joined together by interconnecting wires. (Photos.com)

rotating and/or flipping a mattress twice a year helps to more equally distribute wear and tear over the course of the mattress's life. The actual technique of mattress flipping has been the subject of some discussion for years because the interval is great enough that it is difficult to remember in which direction the mattress was last flipped or rotated without making some kind of mark on the mattress as a reminder.

A mattress can be rotated along three orthogonal axes (x, y, and z); or to compare a mattress to an airplane, roll, pitch, and yaw. The roll axis parallels the longest dimension, the pitch the next-longest, the yaw the shortest. Because a mattress has two sides suitable for sleeping on, and each of those sides has two possible orientations, this means that there are four possible mattress configurations. One mattress-flipping technique that cycles through these four configurations is called the "Klein 4-group," named for mathematician Felix Klein, which is a group describing the symmetries of a rectangle in three-dimensional space. Absent a mnemonic device to remember the previous and next configuration, random selection may be the best choice to maximize the efficiency of mattress flipping. Over the course of 10 years of random selection every six months, for instance, the most-used orientation will be used about 31% of the time, and the least-used about 19%—a 6% deviation from perfectly distributed usage.

Further Reading

Hayes, Brian. *Group Theory in the Bedroom, and Other Mathematical Diversions*. New York: Hill and Wang, 2008.

Krasny, John, William J. Parker, and Vytenis Babrauskas. *Fire Behavior of Upholstered Furniture and Mattresses*. Norwich, NY: Noyes Publications, 2001.

Bill Kte'pi

Microwave Ovens

Category: Architecture and Engineering.
Fields of Study: Algebra; Geometry; Measurement.
Summary: An accidental discovery led to the use of microwave ovens for cooking, a process that continues to be studied.

In 1873, James Clerk Maxwell, using only mathematical considerations, formulated the electromagnetic theory. Maxwell's equations are fundamental to physics and engineering and describe light as a form of electric and magnetic energy. Fifteen years later, experiments carried out by Heinrich Hertz validated Maxwell's theory of electromagnetic waves. This development is a good example of mathematics as a creative medium for the development of science and technology. One of the technological products of Maxwell's theory can be found in most homes in developed countries. Domestic microwave ovens have become increasingly popular since the 1960s, as the device offers a quick method for heating food compared to conventional heating methods. The discovery of electromagnetic waves by Maxwell shows how pure abstract mathematics can generate new technologies. Applied mathematicians also learn new mathematics from problems motivated by this type of application.

Electromagnetic Waves

Electromagnetic waves are a form of radiation represented by their frequency and wavelength. Frequency is the number of cycles that occur in a second and is measured in Hertz (Hz). Wavelength is the measure of the distance over which the wave's shape repeats (λ). The electromagnetic spectrum consists of all possible frequencies and wavelengths of electromagnetic radiation, for example, radio waves, microwaves, infrared, visible, ultraviolet, X-rays, and gamma rays. Microwaves are electromagnetic waves with high frequencies (between 300 MHz and 300 GHz and short wavelengths (from as long as one meter to as short as one millimeter). Besides microwave ovens, practical applications of microwave technology can be found in cellular telephones, radar, satellites, and medical systems.

Discovery

The discovery that microwaves could be used for heating food is one of the accidental cases in the history of science. It occurred in 1945 when Percy Spencer, an American self-taught engineer, was working with microwaves in a radar system and a peanut chocolate bar that was in his pocket started to melt. In the same year, after some experiments with popcorn and eggs, Spencer created the microwave oven. It consisted of a metal box with a high-density electromagnetic field to heat food quickly and efficiently. Twenty years later,

Microwave technology can also be used for melting metal, saving energy and reducing cycle time. (Department of Energy)

microwave ovens were adapted for domestic use as the typical consumer microwave ovens that are known today.

How it Works
The physical and operating principles of microwave ovens are quite simple. Most foods are composed of polarized molecules that are bound together in different ways. When microwave radiation is exposed to food, the molecules within the food are forced to align themselves with a rapidly changing alternating electrical field. Charged molecules oscillate and gain thermal energy via friction. Therefore, microwave radiation can heat food when the radiation is absorbed. This process is dependent on the time of radiation exposition, type of food, and the way the radiation is distributed (scattered, reflected, or transmitted).

In the early twenty-first century, mathematicians are working in universities and industries where interesting problems can be solved using a mathematical approach. Industrial mathematicians at the University of Bath have been working on the microwave cooking process.

A problem with this process is that it can result in localized points inside a food where the radiant electromagnetic field is relatively weak—the temperature in this point may be lower—and the food will be poorly cooked. Theoretically, it is possible using a combination of both analytical and numerical calculation to create a three-dimensional field simulation of this process.

Through a mathematical simulation, an averaged electromagnetic field can be calculated, and it will be possible to determine how it penetrated a moist foodstuff. This example from applied mathematics shows us how mathematics can be used to help us create and enjoy the benefits of technology.

Further Reading
Budd, Chris. "Confessions of an Industrial Mathematician." http://www.math.leidenuniv .nl/~naw/serie5/deel09/jun2008/budd.pdf.

Gallawa, J. Carlton "A Brief History of the Microwave Oven." Southwest Museum of Engineering, Communications and Computation. http://www .smecc.org/microwave_oven.htm.

University of Colorado. "How Microwaves and Microwave Ovens Work." Physics 2000, Einstein's Legacy. http://www.colorado.edu/physics/2000/ microwaves.

Maria Elizete Kunkel

Movies, Making of

Category: Arts, Music, and Entertainment.
Fields of Study: Algebra; Geometry; Representations.
Summary: A variety of mathematics, including signal processing, geometry, and lighting, are required for making movies.

It takes many people with many different talents to make a movie. Some of these required talents are very technical, so these filmmakers must have a working knowledge of various mathematical principles to employ the tools of their trade. A sampling of these areas includes camerawork, sound recording, and special effects. Signal processing, a branch of applied mathematics, is necessary both during the production of a film (for selection of filters and set dressings of acceptable visual frequency) and in postproduction, where dialogue must be made understandable in the sound track. In addition, often the shooting of a scene itself, with its restrictions on space and desired camera angles as well as satisfying lighting needs, becomes a problem in geometry. Physical phenomena and their interactions can increasingly be modeled using mathematics. Mathematicians such as Tony DeRose, who won a 2006 scientific and technical Academy Award for his work on surface representations, play an increasingly important role in producing modern special effects.

Camera Work

In the shooting of a scene, the number of variables is considerable, and those directing the operation of a camera have many decisions to make. Considerations include viewing angles, shutter speeds, lens selection, current lighting, and the format of the film. These considerations become considerably more complicated when working with miniatures in which an attempt is made to fool the eye of the viewer into believing the miniature is a real, full-sized object. The choice of the camera itself, which has many parameters, has a significant effect on the look of the film.

The operation of the camera depends a great deal on the lighting of the set. An f-stop, which has been used for many years on cameras, is the ratio of the focal length of the lens to the diameter of the entrance pupil. This unit was used to control the quantity of light reaching the film. However, because of the fact that much of the light reaching the film plane is lost to diffraction, reflection, and refraction, more modern cameras use the T-stop calibration, which is a measure of the actual amount of light reaching the film plane. If no light were lost to optical factors, these two values would be identical. Both of these measures are used extensively: the f-stop for depth of field calculations and the T-stop for light transmission.

The gaffer (crew boss responsible for planning the lighting) uses a variety of tools to light a scene so that the film can be recorded with the desired viewing window, shutter speed, and camera angles, as well as various aesthetic considerations. One such tool is the inverse square law. This law states that the intensity of a single source of light decreases in proportion to the square of its distance from the subject. Using this law, a small light puts less light on the background, if desired, or a larger light farther away creates a larger area with a similar light level. The light used will also affect the T-stop to be used on the camera, so the light placements must be planned carefully and light output levels must be known exactly.

One calculation the camera operator must constantly make is to determine the depth of field. A lens can focus on only one distance at a time. Therefore, technically, both the foreground and the background of a scene are never in focus simultaneously; in fact, only one point on an actor is in focus at any one time. However, objects close to this distance will not appear blurry to the human eye, which does not perceive imperfection within a certain distance of the point of focus. The distance interval in which all objects are acceptably focused is called the "depth of field." To determine the depth of field, one must first determine the hyperfocal distance, the smallest distance such that all objects from half this distance through infinity are in acceptable focus. This distance can be approximated algebraically, with a parameter known as the "circle of confusion" determining what is considered to be acceptable focus dependent on the focal length and f-stop setting of the lens. Finally, the near and far limits of the depth of field can be determined with the equations

$$\frac{1}{D_n} = \frac{1}{S} + \frac{1}{H} \text{ and } \frac{1}{D_f} = \frac{1}{S} - \frac{1}{H}$$

where D_n and D_f are the near and far limits of the depth of field, S is the distance from the camera to the subject, and H is the hyperfocal distance. These formulas are simplified versions of the normal depth of field equations, which have an interesting geometric derivation.

Audio and Visual Signal Processing

The production sound mixer is in charge of recording the sound and dialogue for a film. Typically, crewmembers are hired to operate microphones, often using long poles with a microphone on the end. These microphones are used to record various sounds on the set, with wireless microphones attached to the actors to record dialogue. Sound effects are recorded separately, as is the score. During post-production, unwanted noise must be filtered out of the recordings, the dialogue must be made understandable, and the effects, score, and dialogue must be mixed together meaningfully.

To remove background noise, the audio signal (composed of sound waves) is decomposed using a Fourier transform, so that the model of the audio signal is divided into simpler, trigonometric components. These components are then analyzed, isolating frequencies corresponding to unwanted artifacts in the recording, such as the sound of the wind on the microphone. Background noise is removed by removing the Fourier components of amplitude below a certain level. Finally, by reversing the transform, a more filmworthy audio signal is obtained.

Processing must also be done to the visual signal. Video cameras record at a "frame rate," the frequency with which the camera produces images. These images are recorded as discrete signals, which are then reconstructed on film. If objects of a high visual frequency are used in a scene, a loud tie for example, then the image on film will experience aliasing, causing visual distortion or artifacts. To avoid this, the set designer or costumer needs to avoid objects above a certain visual frequency. This frequency, called the "Nyquist frequency," is half the frame rate. If images of high-visual-frequency objects are desired, then an antialiasing filter must be used, such as a lowpass filter, which will pass the low-frequency objects but reduce the amplitude of the high-frequency objects. Filmmakers have many filters that can be used to capture a wide variety of objects in a scene, depending on the mix of visual frequencies present.

Further Reading

Burum, Stephen, ed. *American Cinematographer Manual.* Hollywood, CA: ASC Press, 2007.

Haunsperger, Deanna, and Steve Kennedy. "Math Makes the Movies." *Math Horizons* 9 (November 2001).

McAdams, A., S. Osher, and J. Teran. "Crashing Waves, Awesome Explosions, Turbulent Smoke, and Beyond: Applied Mathematics and Scientific Computing in the Visual Effects Industry." *Notices of the American Mathematical Society* 57, no. 5 (2010).

WILLIAM GRIFFITHS

MP3 Players

Category: Communication and Computers.
Fields of Study: Algebra; Communication; Data Analysis and Probability; Number and Operations.
Summary: Mathematics and mathematical data compression algorithms make MP3 players possible.

MP3 players have revolutionized the way people listen to music. MPEG Audio Layer III (MP3) is an audio compression standard that reduces music files with little perceptible loss of quality. It is one of the Motion Pictures Expert Group standards for lossy compression. The inventors of MP3, according to the United States MP3 patent, are engineers Bernhard Grill, Karl-Heinz Brandenburg, and Bernd Kurten; computer scientist Thomas Sporer; and mathematician Ernst Eberlein. The development was mathematically and technically challenging according to Brandenburg, who is sometimes called a specialist in mathematics and electronics. He stated, "in 1991, the project almost died. During modification tests, the encoding simply did not want to work properly. Two days before submission of the first version . . . we found the compiler error." Scientists at Fraunhofer-Gesellshaft developed an MP3 player in the early 1990s.

In 1997, engineer Tomislav Uzelac invented the AMP MP3 Playback Engine, which is regarded as the first successful MP3 player. Computer science student Justin Frankel, who also helped develop the peer-to-peer Gnutella network, and fellow student Dmitry Boldyrev created the free MP3 player Winamp in 1998. Inventor Briton Kramer contributed to the first mass-produced player MPMan. The ability to share files over the Internet, legally and illegally, for free or for purchase, was a significant factor in the rapid spread of the MP3 format. By the twenty-first century, iPods became one of the most popular MP3 players, in part because of the availability of music and video via the iTunes store. The ability to hold thousands of songs, videos, and other types of files is one of the benefits of MP3 players, all of which would not be possible without mathematics and mathematical data compression algorithms.

Compression and Encoding

Data compression is either "lossy" or "lossless," referring to whether any data is discarded in the process of creating a smaller file. Huffman coding, developed by mathematician David Huffman, is used for MP3 compression. It employs a mathematical idea called a "frequency-sorted binary tree" to look for recurring strings of binary information in the digital file. These strings are replaced by shorter binary codes. The most frequently occurring strings are assigned the shortest replacement codes, optimizing compression. In lossless compression, all original information is preserved in some way. In lossy compression, some information is discarded to decrease file size. MP3 compression relies, in part, on perceptual coding.

In a human ear, certain waveforms are indistinguishable. Psychoacoustic models prioritize data according to the ear's ability to distinguish the sounds the data produce. Mathematical models of auditory processing yield encoding information and algorithms, such as frequency threshold curves, masking functions, and critical bandwidths. Signal processing typically relies on Fourier transforms, named for mathematician Joseph Fourier, to enable coding and decoding. Ultimately, an MP3 music file consists of a series of short, dependent frames, like a filmstrip. Each frame has a header with information about the data in the frame. Inside the frame is audio information in frequencies and amplitudes. Sometimes, at the beginning or end, there is an ID3 data block, which stores the artist name, track title, album name, recording year, or other information.

Sound

Optimization of audio playback depends not just on the human ear but on the equipment used. Speakers are common on computers, while most MP3 players use some form of over-ear headphones or earbuds that fit into the ear canal. Empirical studies suggest that noise from internal earbuds may be damaging to hearing because the decibel level experienced by listeners is higher on average than with external earphones, and long-life batteries players allow people to listen longer. Some researchers have reported average listening levels of about 110–120 decibels, equivalent to a rock concert. Based on findings of such studies, many audiologists recommend using noise-canceling headphones rather than turning up the volume. Engineer Lawrence Fogel first explored noise-canceling headphones for aviation in the 1950s. Noise cancellation uses the mathematical properties of waves to create a signal with the same amplitude but with an inverted phase to unwanted noise, creating a combined wave inaudible to the human ear.

Shuffle

One other interesting mathematical problem related to MP3 players is the shuffle function. Various mathematical algorithms are used to permute the play order of songs in an MP3 player's library. In the early twenty-first century, the iPod's default shuffle system reorders songs much like someone shuffling a deck of cards, giving each song an equal chance to end up in any position in the shuffle and resulting in no repeats. However, many factors can affect perceived randomness and equal likelihood of orderings. For example, users can request higher chances of play for songs with high user ratings. Songs can also be marked "Skip When Shuffling" so that they are completely excluded. Most people frequently reshuffle, generating new random orderings before completing the library, and so some tracks appear to repeat or group in nonrandom ways.

Further Reading

Kallen, Stuart. *iPods and MP3 Players*. Florence, KY: Gale Cengage, 2010.

Salomon, David. *A Guide to Data Compression Methods*. New York: Springer, 2002.

BILL KTE'PI

Neural Networks

Category: Communication and Computers.
Fields of Study: Data Analysis and Probability; Number and Operations; Representations.
Summary: Artificial neural networks use sophisticated mathematical algorithms and computational functions to simulate biological neural networks.

The term "neural networks" is generally applied to the systems of biological or artificial neurons. More often it is used in application to artificial neural networks that are designed to reproduce some human brain functions, such as information processing, memory, and pattern recognition. However, this term is also used for biological neural networks, for which the term "neural

Figures 1A and 1B. Biological and Artificial Neurons.

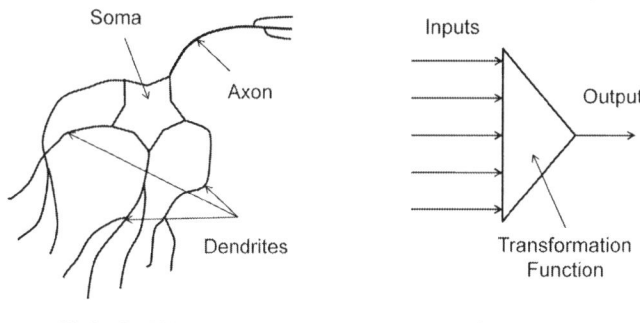

system" is more common. The beginning of modern neural network research is credited to neuroscientist Warren McCulloch and mathematician Walter Pitts in 1943. McCulloch had spent decades pondering the logic of the nervous system (for example, what allows people to think or feel) before beginning his collaboration with Pitts. He specifically credited Pitts's knowledge of modular arithmetic for the success of their joint work, which produced the McCulloch–Pitts Theory of Formal Neural Networks. Their research suggests that any computable function can be completely realized by a McCulloch–Pitts artificial neural network, though some such networks would be impractically large.

Artificial Neural Networks

Artificial neural networks are mathematical tools or physical devices that function similarly to biological neural systems. They consist of building blocks, called "artificial neurons," which resemble the structure of real neurons. Each biological neuron includes three major parts: dendrites, soma, and axon (see Figure 1A). Correspondingly, each artificial neuron also consists of three major parts: inputs (or "dendrites"), transformation function ("soma"), and output ("axon") (see Figure 1B). The terminology that is generally used for biological neurons is also often applied to artificial neurons.

Modern neural networks use data analysis and non-linear statistical methods to model complex relationships between inputs and outputs or to find patterns. Bayesian methods of inference, named for Thomas Bayes, are increasingly employed. Graph theory and geometry are also very useful for mapping neural networks, assessing their capabilities, and studying pattern classification. Artificial neural networks are applied to a variety of problems in science, industry, and finance in which people must draw conclusions and make decisions from noisy and incomplete data. They can perform pattern recognition and pattern classification, time series analysis and prediction, function approximation, and signal processing. Several types of artificial neural networks were developed for the specific problems for which they can find the best solution. The most famous of them are single- and multi-layer perceptrons; Hopfield neural networks, named for John Hopfield; self-organizing Kohonen maps, named for Tuevo Kohonen; and Boltzmann machines, named for the Ludwig Boltzmann distribution. Regardless of the type of neural network or the problem it is designed to solve, the output is some mathematical function of the inputs, often involving probability distributions. As examples, consider functions of the three types of the artificial neural networks represented in Figure 2.

The first, single-layer perceptron consists of one layer of artificial neurons and was designed for pattern recognition and classification problems (see Figure 2A). The input pattern of signals s_i is fed to each neuron in the perceptron with different weights, w_{ij}. Then the signals are added in each jth neuron to form a weighted sum $\Sigma_i w_{ij} s_i$, which is processed by a transformation (nonlinear) function, resulting in a pattern of the output signals o_j. Thus, the pattern of output signals o_j is determined by the set of weights w_{ij}, and this set of weights forms a memory in the neural network. To

Figures 2A, 2B and 2C. Artificial Neural Networks.

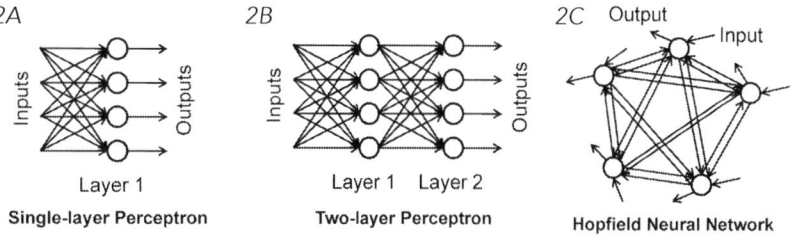

obtain desired response pattern d_j to a given input pattern s_i, the perceptron is required to be "trained." Training (or learning) procedure consists of the method that adjusts neural network weights w_{ij} that form desired output pattern d_j.

Because of limited capability of single-layer perceptrons (for example, they cannot reproduce "exclusive OR" logical operations), the multilayer perceptrons (see Figure 2B) became very popular for different problems in pattern recognition and classification. Inclusion of one or more "hidden" layers into the neural networks increased their learning capability and performance. Multilayer perceptrons are learned by so-called back-propagation algorithm that changes weights w_{ij} in all layers to ensure desired output in the last layer.

Both single-layer and multilayer perceptrons belong to a class of feedforward neural networks, as connections between the neurons do not form closed loops (see Figures 2A and 2B), and information transfers only in one direction, from the input to the output. A Hopfield neural network is a representative of another class, recurrent artificial neural networks, with bidirectional flow of information (see Figure 2C). Each neuron in this network is connected to the others with symmetric bidirectional connections, and its output is calculated in a way similar to that for perceptrons. A Hopfield neural network runs by cycles. During one cycle, the output of each neuron is calculated using external inputs and neural outputs from the previous cycle. These neuronal outputs become their inputs, with corresponding weights and transformation function, during the next cycle. Neural outputs are recalculated for each cycle until the system reaches a steady state. This steady state pattern of neural outputs represents a stored pattern in the Hopfield neural network. Information in Hopfield neural networks, as in perceptrons, is stored in the weights, w_{ij}.

Further Reading

Coolen, A. C. C. "A Beginner's Guide to the Mathematics of Neural Networks." In *Concepts for Neural Networks*. Edited by L. J. Landau and J. G. Taylor. New York: Springer, 1998.

Haykin, Simon. *Neural Networks: A Comprehensive Foundation*. 2nd ed. Upper Saddle River, NJ: Prentice Hall, 1999.

Kandel, Eric, James Schwartz, and Thomas Jessell. *Principles of Neural Science*. 4th ed. New York: McGraw-Hill, 2000.

Rumelhart, David E., James L. McClelland, and the PDP Research Group. *Parallel Distributed Processing*. Vol. 1. *Foundations*. Cambridge, MA: MIT Press, 1986.

Vladimir E. Bondarenko

Nielsen Ratings

Category: Arts, Music, and Entertainment.
Fields of Study: Communication; Data Analysis and Probability; Measurement.
Summary: Television viewing data are estimated using metrics collection and statistical modeling.

The Nielsen Ratings are a measure of how many people are watching certain television programs. When Arthur Nielsen began measuring television viewing in 1950, there were three networks and about 9% of U.S. households had a television. In the twenty-first century, homes often have multiple televisions receiving scores of channels. Even when the number of television sets was small, it was not possible to gather complete viewing data for every single person who owns a television. Instead, Nielsen Media Research uses statistical sampling methods to take a representative subset of viewers and then extrapolates from the sample's viewing activities to the whole population of viewers. The statistical methods Nielsen uses to collect its data have been refined several times in response to changes in viewer behavior. People who develop and analyze ratings typically have expertise in analytics, metrics, and statistical modeling.

Advanced statistical methodologies, like data mining and software such as Mathematica, are used to extract patterns from Nielsen data that help explain which segments of the population view particular shows. Networks make decisions about whether to cancel or renew programs based on Nielsen ratings. Companies also use Nielsen audience estimates to allocate tens of billions of television advertising dollars each year.

Statistical Sampling and Data Collection

No one knows exactly how many households have televisions, but 2010 estimates suggest that the average U.S. household has just fewer than three televisions. Using statistical sampling, Nielsen can obtain representative data using small a proportion of households: approximately 9000 in its national sample, another 1000 in its Hispanic sample, and various smaller amounts in selected local markets. For the national television sample and major local markets, "people meters" record what television shows household members watch using an electronic set meter, along with a remote control that distinguishes each individual member of the household. Set meters are also used to collect data in mid-sized local markets, but with paper diaries for individual demographics. Meters transmit data to Nielsen every night, where it is checked mathematically for transmission or recording errors before analysis. In the smallest markets, viewers record programs in paper diaries and mail them to Nielsen. Historically, Nielsen tracked only television programs that were viewed live at the time they aired. However, people are increasingly using digital video recorders (DVRs), streaming video, and other delayed viewing technologies, which biases live ratings and affects both programming and advertising decisions. Nielsen began adding DVR households to its sample in 2006 and now regularly reports same-day and seven-day DVR playback ratings as well as its traditional live viewer ratings. People's failure to return paper diaries is also a growing source of bias, and research methodologists are working on revising this method to make completing the diaries easier to encourage greater response.

Television Metrics

Nielsen's primary metrics for television viewing are rating, share, and projected audience. A program's rating is a percentage that represents the number of households that watched the program out of the total number of households that could have watched the program. In this case, the denominator of the fraction is fixed according to the Nielsen sample size. The 1983 finale of the television show *M*A*S*H* holds the record for highest Nielsen rating, 60.2, which means slightly more than 60% of possible sample households tuned in to watch. At the time there were about 83 million television households, so one sample rating point represented 1% or 830,000 households in the population. However, it would be very unusual for every household to be watching television at the same time.

Share adjusts for this fact by computing the percentage of households that watched a specific program out of the number of households that were actually watching television during that time. This is a more complicated calculation, since the number of televisions being used at any given moment changes constantly. Shares are often used to measure how competitive a program is in its particular time slot. The *M*A*S*H* finale had a 77 share, which means 77% of households watching television at all were tuned to that program. Ratings and shares are also computed for several age, race, and other subgroups, as these are very important to advertisers. Since data are recorded at the household level—and many people may watch the same program in one house, or outside the home in places like sports bars or dorms—the number of individual viewers in a subgroup or population can only be estimated from the demographic data recorded by people meters and diaries. In 2007, Nielsen also began to measure college students' viewing habits by treating them as if they were watching an additional television set at home.

Projected audience is the estimated number of people reached in the overall population, which is calculated using statistical modeling. The *M*A*S*H* finale had a projected audience of 106 million viewers. The number of television households grows every year, as does the number of channel choices, so it can be difficult to compare ratings from across years, especially over large stretches of time. For example, although Super Bowl XLIV surpassed the *M*A*S*H* finale in terms of estimated viewers (106.5 million), it had a lower rating (46.4). Another reason that the numbers may be difficult to compare is that Nielsen produces rapid overnight ratings for many media outlets and these values are later adjusted. Further, only selected numbers are made public, such as the daily or weekly top 20 shows, which vary from week to week. Comprehensive data is generally available only to Nielsen's clients, and networks may advertise only the statistics that are most favorable.

Further Reading

Balnaves, Mark, Tom O'Regan, and Ben Goldsmith.
Rating the Audience: The Business of Media. New York: Bloomsbury, 2011.

Nielsen Company. "How DVRs are Changing the Television Landscape." http://blog.nielsen.com/nielsenwire/wp-content/uploads/2009/04/dvr_tv landscape_043009.pdf.

Webster, James, Patricia Phalen, and Lawrence Lichty. *Ratings Analysis: The Theory And Practice Of Audience Research*. 3rd ed. Mahway, NJ: Lawrence Erlbaum Associates, 2008.

Carmen M. Latterell

Optical Illusions

Category: Arts, Music, and Entertainment.
Fields of Study: Geometry; Measurement; Problem Solving.
Summary: Optical illusions are predictable illusory phenomena that are not yet fully understood.

Optics, generally, is the science of the visible. Physical optics is the study of the nature and propagation of light. Physiological optics is the study of neurophysiological processes of light reception and image forming as conditions of vision and merges with psychology and cognitive science into a unitary "vision science." Optical illusions likely have been observed as long as mankind has existed. Some optical illusions arise from people's ability to see in three dimensions, even though retinal images are flat representations on a curved surface. Extracting three-dimensional information from ambiguous two-dimensional images requires interpretive rules in the brain. Many optical illusions have mathematical connections, especially in the perception of geometry within the illusion. They are popular as entertainment, and mathematics teachers sometimes use optical illusions in the classroom in order to engage students and to develop visualization skills.

Examples

Physical phenomena leading to seeing unreal things, or to seeing real things in a distorted way (for example, phenomena due to special atmospheric conditions: halos, coronas, and sightings of distant objects caused by reflections between air layers of different density) are now well understood, and not usually named "illusions." The apparent flattening of the sun disc at the sunset is in accord with the laws of light propagation (differential refraction), but it is not an illusion. Illusions of perception are situations "when perception goes wrong" and where a central (neurophysiological or psychological) cause must be supposed—something is perceived as something else (error of identification) or is perceived differently than it is (error of quality or quantity). For example, the moon at the horizon is often reported to appear larger than if seen high in the sky, although the angular size of the moon disc is in both instances the same (approximately 30 arc minutes)—this is the famous "moon illusion."

Illusory phenomena have been observed since ancient times, for example, the "moon illusion" was known to Ptolemy, and an illusory "motion aftereffect" caused by watching a waterfall was mentioned by Aristotle. However, the proper scientific study of visual illusions began in the middle of the nineteenth century with the discovery of geometric-optical illusions, (distortions of perceived lengths, sizes, and shapes observed in simple drawings or in real-world situations). For example, a path in the visual field subdivided into a series of segments usually appears longer than the same path that is empty (see the Oppel–Kundt illusion, Figure 1). Lengths of linear segments may be overestimated or underestimated, depending on added elements (for example, the popular Müller–Lyer illusion). Geometric figures drawn over linear or curvilinear rasters often appear deformed (see the Hering, Zöllner, or Ehrenstein–Orbison illusion Figure 2). Other instances of optical illusions involve judgments of brightness (see Figure 3), and particularly illusory "contrast phenomena," such as well-known Mach bands, or the Hermann grid (see Figure 4). More recently, dynamic phenomena, such as illusory motion seen in static pictures, or the famous "scintillating grid," have been described.

Generally, all these phenomena demonstrate the universal principle of context-dependence in visual (and any) perception: a stimulus, S, is perceived differently if presented together with a context stimulus S' than if presented alone. In other words, a purely attentional separation of S from S' is impossible in spite of the observer's effort.

What differentiates all these phenomena from incidental errors of perception is that they occur regularly and predictably in most or all observers. After more than 150 years since their discovery, there is still no satisfactory theory of these phenomena, although a great variety of explanations have been proposed.

Explanations

The two main directions of explanatory approaches have traditionally been the empiricist and the nativist theories. The empiricist theories, going back to Herman Helmholtz's theory of unconscious inferences, emphasized the role of the subject's past experience and of cognitive factors forming the perception. By contrast, the nativist theories searched for explanations in the structure and the functional principles of the sensory organ itself.

Empiricist theories, in spite of their speculative character, have been revitalized by cognitive psychologists and are still influential; for example, a popular theory sought to explain a group of optical illusions as results of inappropriate constancy scaling due to erroneous perspectival interpretation of the illusion-inducing figure. However, these theories ignore much of empirical counter evidence, such as tactile analogies of certain optical illusions, or geometrico-optical distor-

Figures 1 and 2

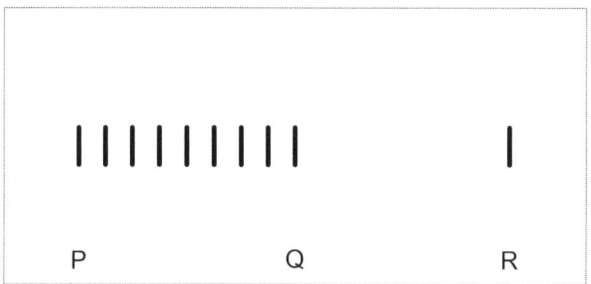

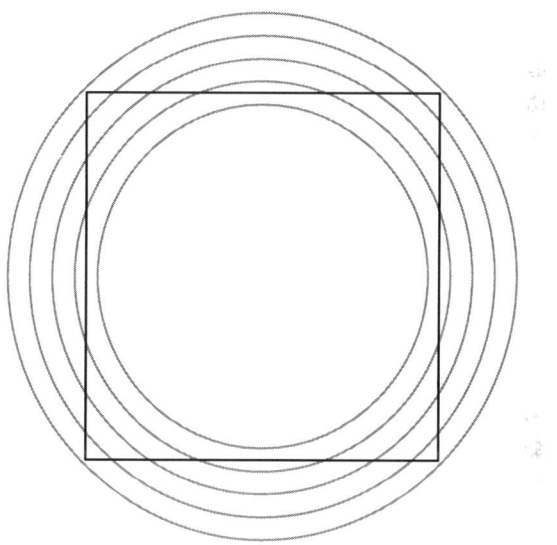

Figure 1. Oppel-Kundt illusion: the length of the PQ segment appears greater than the segment QR, although PQ = QR.

Figure 2. Ehrenstein-Orbison illusion: sides of the square drawn in an array of concentric circles appear inward-bent, although they are really segments of straight lines.

Figures 3 and 4

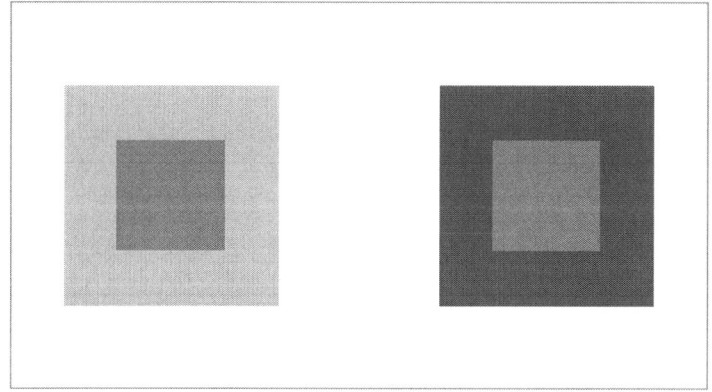

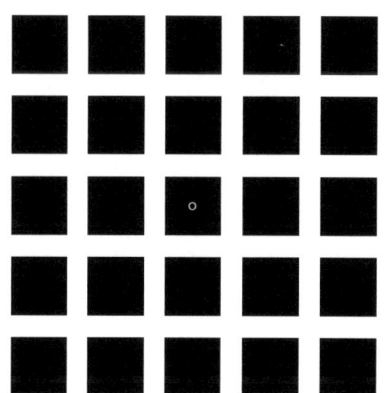

Figure 3. Context effects on perceived brightness: the inner square of the left-hand-side figure appears darker than the inner square of the right-hand-side figure, although they are printed with exactly the same gray-shade level. Figure 4. Hermann grid: illusory grayish shadows are seen at the crossings of white stripes, although objectively the background is uniformly white. (Fixate on the figure center marked by the circle.)

tions observed in contexts not suggesting any perspectival interpretation. Neonativist theories integrating approaches of Gestalt psychology and neurophysiology and searching for interactions within higher levels of the visual system are arguably more promising, although they are usually limited to circumscribed groups of illusory phenomena. The general opinion in the early twenty-first century is that the broad variety of optical illusions cannot be explained by a single cause; therefore, a unitary theory of optical illusions is rather unlikely.

Optical illusions are neither deceptions of the eye nor errors of the cognitive processing of sensory data. They are facts of vision, presumably manifestations of the functional principles of the visual system in its entirety. The same functional principles, or the "laws of seeing," are at work in visual arts, or in visualization technologies such as virtual reality. The study of optical illusions in laboratory as well as in natural environments importantly contributes to the understanding of the process of vision and of the nature of the visual life world.

Further Reading

Boring, E. G. *Sensation and Perception in the History of the Experimental Psychology.* New York: Appleton-Century-Croft, 1942.

Coren, S., and J. S. Girgus. *Seeing Is Deceiving. The Psychology of Visual Illusions.* Hillsdale, NJ: Lawrence Erlbaum, 1978.

Gombrich, E. H. *Art and Illusion.* Oxford, England: Phaidon Press, 1977.

Metzger, W. *The Laws of Seeing.* Cambridge, MA: MIT Press, 2006.

Robinson, J. O. *The Psychology of Visual Illusion.* 2nd ed. New York: Dover, 1998.

Ross, H. E., and C. Plug. *The Mystery of the Moon Illusion.* New York: Oxford University Press, 2002.

Seckel, Al. *Masters of Deception: Escher, Dali & the Artists of Optical Illusion.* New York: Sterling, 2007.

———. *Optical Illusions: The Science of Visual Perception.* Buffalo, NY: Firefly Books, 2009.

Jiri Wackermann

Parallel Processing

Category: Communication and Computers.
Fields of Study: Algebra; Number and Operations.
Summary: Parallel processing speeds up the run-time of computing through the use of mathematical algorithms.

In computing, parallel processing is the action of performing multiple operations or tasks simultaneously by two or more processing cores. Ideally, this arrangement reduces the overall run-time of a computer program because the workload is shared among a number of engines—central processing units (CPUs) or cores. In practice, it is often difficult to distribute the instructions of a program in such a way that each CPU core operates continuously and efficiently, and without interfering with other cores. It should be noted that parallel processing differs from multitasking, in which a single CPU core provides the effect of simultaneously executing instructions from several different programs by rapidly switching between them, or interleaving their instructions. Modern computers typically include multi-core processor chips with two or four cores. The most advanced supercomputers in the early twenty-first century may have thousands of multi-core CPU nodes organized as a cluster of single processor computers and connected using a special-purpose, high-speed, fiber communication network. Although it is also possible to perform parallel processing by connecting computers together using a local area network, or even across the Internet, this type of parallel processing requires the individual processing elements to work predominantly in isolation because of the comparatively slow communication between nodes. Parallel processing requires data to be shared among processors and thus leads to the concept of "shared memory" where multiple processing cores work with the same physical memory. In large computer clusters, the memory is usually distributed across the nodes, with each node storing its own part of the full problem. Data are exchanged between nodes using message-passing software, such as Message Passing Interface (MPI).

Amdahl's Law and Gustafson's Law

The speed-up gained through parallelization of a program would ideally be linear; for example, doubling the number of processing elements should halve the run-time. However, very few parallel algorithms achieve this target. The majority of parallel programs attain a near-linear speed-up for small numbers of processing elements but for large numbers of processors the addition of further cores provides negligible benefits.

The potential speed-up of an algorithm on a parallel computing platform is given by Amdahl's law, originally formulated by Gene Amdahl in the 1960s. A large mathematical or engineering problem will typically consist of several parallelizable parts and several non-parallelizable parts. The overall speed-up attainable through parallelization is proportional to the size of the non-parallelizable portion of the program and is given by the equation

$$S = \frac{1}{1-P}$$

where S is the speed-up of the program (as a factor of its original sequential runtime), and P is the fraction that is parallelizable. Amdahl's law assumes the size of the problem is fixed and that the relative proportion of the sequential section is independent of the number of processors. For example, if the sequential portion of a program is 10% of the run-time ($P = 0.9$), no more than a 10-times speed-up could be obtained, regardless of how many processors are added. This characteristic puts an upper limit on the usefulness of adding more parallel execution units.

Gustafson's law is closely related to Amdahl's law, but is not so restrictive on the assumptions made about the problem. It can be formulated algebraically as

$$S(P) = P - a(P-1)$$

where P is the number of processors, S is the speed-up, and a is the non-parallelizable proportion of the process.

Applications

Parallel computing is used in a broad range of fields, including mathematics, engineering, meteorology, bioinformatics, economics, and finance. However, all of these applications usually involve performing one or more of a small set of highly parallelizable operations, such as sparse or dense linear algebra, spectral methods, n-body problems, or Monte Carlo simulations. Frequently, the first step to exploiting the power of parallel processing is to express a problem in terms of these basic parallelizable building blocks.

Parallel processing plays a large part in many aspects of everyday life, such as weather prediction, stock market prediction, and the design of cars and aircraft. As parallel computers become larger and faster, it becomes feasible to solve larger problems that previously took too long to run on a single computer.

Further Reading

Barney, Blaise. "Introduction to Parallel Computing." *Lawrence Livermore National Laboratory*, 2007. https://computing.llnl.gov/tutorials/parallel_comp/.

Gupta, A., A. Grama, G. Karypis, and V. Kumar. *An Introduction to Parallel Computing: Design and Analysis of Algorithms*. Reading, MA: Addison Wesley, 2003.

Jordan, Harry F., and Gita Alaghband. *Fundamentals of Parallel Processing*. Upper Saddle River, NJ: Prentice Hall, 2002.

Chris D. Cantwell

Personal Computers

Category: Communication and Computers.
Field of Study: Algebra; Communication; Data Analysis and Probability; Number and Operations; Representations.
Summary: Advances in computing have made mathematical processing power so inexpensive that it has become more practical to do many tasks on the computer.

A computer is a device that manipulates raw data into potentially useful information. Computers may be analog or electronic. Analog computers use mechanical elements to perform functions. For example, Stonehenge in England is believed by some to be an analog computer. It allegedly uses the stones along with the positions of the sun and moon to predict celestial events like the solstices and eclipses. Electronic computers use electrical components like transistors for computations.

Many consider the first personal computer to be Sphere 1, created by Michael Wise in the mid-1970s.

The Apple II was introduced in 1977, and Apple Inc. offered the Macintosh, which had the first mass-marketed graphical user interface, by 1984. IBM debuted its personal computer in 1981. "Macs" and PCs quickly became common in businesses and schools for a variety of purposes. Processing speed, size, memory capacity, and other functional components have become faster, smaller, lighter, and cheaper over time, and personal computers have evolved into a multitude of forms designed to be customizable to each user's needs. At the beginning of the twenty-first century, desktops, laptops, netbooks, tablet PCs, palm-sized smartphones, handheld programmable calculators, digital book readers, and devices like Apple's iPad offer access to computing, the Internet, and other functions.

Mathematical History of Computers

Modern computing can be traced to nineteenth century mathematician Charles Babbage's analytical engine. Boolean algebra, devised by mathematician George Boole later in the same century, provided a logical basis for digital electronics. Lambda calculus, developed by mathematician Alonso Church in the early twentieth century, also laid the foundations for computer science, while the Turing machine, a theoretical representation of computing developed by mathematician Alan Turing, essentially modeled computers before they could be built. In the 1940s, mathematicians Norbert Wiener and Claude Shannon researched information control theory, further advancing the design of digital circuits. The Electrical Numerical Integrator and Calculator (ENIAC) was the first general purpose electronic computer. It was created shortly after World War II by physicist-engineer John Mauchly and engineer J. Presper Eckert. They also developed the Binary Automatic Computer (BINAC), the first dual-processor computer, which stored information on magnetic tape rather than punch cards, and the first commercial computer, Universal Automatic Computer (UNIVAC). Mathematician John Von Neumann made important modifications to ENIAC, including serial operations to facilitate mathematical calculations. Scientists William Bradford Shockley, John Bardeen, and Walter Brattain won the 1956 Nobel Prize in Physics for transistor and semiconductor research, which influenced the development of most subsequent electronic devices, including personal computers. During the latter half of the twentieth century, countless mathematicians, computer scientists, engineers, and others advanced the science and technology of personal computers, and research has continued into the twenty-first century. For example, Microsoft co-founder Bill Gates published a paper on sorting pancakes, which has extensions in the area of computer algorithms. Personal computers have facilitated mathematics teaching and research in many areas such as simulation, visualization, and random number generation, though the use of calculators and software like Maple for teaching mathematics generated controversy.

Devices, Memory, and Processor Speeds

The typical personal computer has devices for the input and output of information and a means of retaining programs and data in memory. It also has the means of interacting with programs, data, memory, and devices attached to the computer's central processing unit (CPU). Input devices have historically included a keyboard and a mouse, while newer systems frequently use touch technology, either in the form of a special pad or directly on the screen. Other devices include scanners, digital cameras, and digital recorders. Memory storage devices are classified as "primary memory" or "secondary" devices. The primary memory is comprised of the chips on the board inside the case of the computer. Primary memory comes in two types: read only memory (ROM) and random access memory (RAM). ROM contains the rudimentary part of the operating system, which controls the interaction of the computer components. RAM holds the programs and data while the computer is in use. The most popular types of secondary memory used for desktop computers include magnetic disk drives, optical CD and DVD drives, and USB flash memory.

The speed of the computer operation is an important factor. Computers use a set clock cycle to send the voltage pulses throughout the computer from one component to another. Faster processing enables computers to run larger, more complex programs. The disadvantage is that heat builds up around the processor, caused by electrical resistance. ENIAC was 1000 times faster than the electromechanical computers that preceded it because it relied on vacuum tubes rather than physical switches. Turing made predictions regarding computer speeds in the 1950s, while Moore's law, named for Intel co-founder Gordon Moore, quantified the doubling rate for transistors per square inch on integrated circuits. The number doubled every year from 1958 into the 1960s, according to Moore's data. The rate slowed through the end of the twentieth century to roughly a doubling every 18 months. Some scientists predict more slowdowns because of the heat problem. Others, like mathematician Vernor Vinge, have asserted that exponential technology growth will produce a singularity, or essentially instantaneous progress. Processing speed, memory capacity, pixels in digital images, and other computer capabilities have been limited by this effect. There has also been a disparity in the growth rates of processor speed and memory capacity, known as *memory latency*, which has been addressed in part by mathematical programming techniques, like caching and dynamic optimization.

Carbon nanotubes and magnetic tunnels might be used to produce memory chips that retain data even when a computer is powered down. At the start of the twenty-first century, this approach was being developed with extensive mathematical modeling and physical testing. Other proposed solutions involved biological, optical, or quantum technology. Much of the physics needed for quantum computers exists only in theory, but mathematicians like Peter Shor are already working on the mathematics of quantum programming, which involves ideas like Fourier transforms, periodic sequences, prime numbers, and factorization. Fourier transforms are named for mathematician Jean Fourier.

The Digital Divide

The digital divide is the technology gap between groups that have differential access to personal computers and related technology. The gap is measured both in social metrics, such as soft skills required to participate in online communities, and infrastructure metrics, such as ownership of digital devices. Mathematical methods are used to quantify the digital divide. Comparisons may be made using probability distributions and Lorenz curves, developed by economist Max Lorenz, and measures of dispersion such as the Gini coefficient, developed by statistician Corrado Gini. Researchers have found digital divides among different countries, and within countries, among people of different ages, between genders, and among socioeconomic strata.

The global digital divide quantifies the digital divides among countries and is typically given as the differences among the average numbers of computers per 100 citizens. In the early twenty-first century, this metric varied widely. Several concerted private and government efforts, such as One Laptop Per Child, were directed at reducing the global digital divide by providing computers to poor countries. The breakthroughs connected to these efforts, such as mesh Internet access architecture, benefited all users. The Digital Opportunity Index (DOI) is computed by the United Nations based on 11 metrics of information and communication technologies, such as proportion of households with access to the Internet. It has been found to be positively associated with a country's wealth.

Further Reading

Lauckner, Kurt, and Zenia Bahorski. *The Computer Continuum.* 5th ed. Boston: Pearson, 2009.

Lemke, Donald, and Tod Smith. *Steve Jobs, Steve Wozniak, and the Personal Computer*. Mankato, MN: Capstone Press, 2010.

Wozniak, Steve, and Gina Smith. *iWoz: Computer Geek to Cult Icon: How I Invented the Personal Computer, Co-Founded Apple, and Had Fun Doing It*. New York: W. W. Norton, 2007.

Zenia C. Bahorski
Maria Droujkova

Pulleys

Category: Architecture and Engineering.
Fields of Study: Algebra; Geometry.
Summary: Pulleys provide mechanical advantage and help people do work.

A pulley is a simple machine consisting of a cylinder, called a "drum," "wheel," or "sheave," rotating on an axle, and a rope, chain, or belt running over the cylinder without sliding. Pulley drums often have grooves and ribs that prevent their ropes from sliding over the edge. People use pulleys in three ways: to change directions of forces, to change magnitude of forces, and to transmit power. Pulleys are used in building and construction, ship rigging, and within belt-driven mechanisms.

Mathematicians have investigated many aspects of pulleys. There is evidence that Archimedes of Syracuse used a compound pulley to move a ship and studied the related theories. He famously expressed: "Give me a place to stand and I will move the Earth." While his mechanical inventions brought him recognition among his contemporaries, he seems to have preferred pure mathematics. Guidobaldo del Monte reduced systems of pulleys to levers. Guillaume de l'Hôpital investigated the equilibrium of a pulley system, and mathematicians continue to explore his pulley problem using algebra, geometry, trigonometry, and calculus. A mechanical tide-predicting machine, which incorporated pulleys, is attributed to William Thomson, who later became Lord Kelvin.

Changing Directions of Forces

In an example of this use of pulleys, construction workers often attach pulleys to roofs of buildings. A builder standing on the ground can pull down on one end of the pulley's rope and a weight on the other end will move up as the drum rotates.

The vectors of input and output forces always go along the two ends of the pulley's rope. This means that a pulley can change the direction of a force within the plane that is perpendicular to the pulley's axle but not sideways from that plane. The builder can also stand inside the building, pulling the rope through a window, or on the roof pulling horizontally, as long as the triangle formed by the worker, the weight, and the pulley's drum is perpendicular to the pulley's axle.

Changing Magnitudes of Forces

When a pulley is used to change the magnitude of a force, its axle is attached to the weight, and the pulley moves up together with the weight. For example, a sailor can attach one end of a line to a yardarm, string it around a pulley's drum attached to a weight, and pull the other end up, standing on the yardarm. The sailor will only have to apply the force equal to one-half of the weight.

Does the other half of the force disappear, breaking the conservation of energy law and the work-energy theorem? No, it is distributed to the other, attached end of the rope. Moreover, the sailor will use half the force, but pull enough line to cover twice the distance the weight is lifted. The total work, which is equal to the product of the force and the distance, will be the same as in the fixed pulley case:

$$W = F \times d = \frac{1}{2} F \times 2d.$$

Changing Directions and Magnitudes of Forces: Blocks and Tackles

Because it is much easier to work for longer than to increase one's force, movable pulleys are widely used. A block and tackle is a pulley system where the rope zigzags through movable and fixed pulleys. Depending on the way the tackle is rigged, it can provide a force advantage with the factor of two, as in the example above, or 3, 4, 5 and so on. At first sight, it would seem that a block and tackle can reduce the force required to lift weights by any factor. However, friction interferes increasingly with more pulleys used.

Marine cadets memorize rigging of common block and tackle systems, and the names of tackles corresponding to force advantage factors: factor 2: "gun"; factor 3: "luff"; factor 4: "double"; factor 5: "gyn."

Drums for tackles may have multiple grooves to reduce rope friction. When tackles are combined, for example, a double tackle upon a luff tackle, their force advantage factors multiply, in this case, creating the force advantage of $3 \times 4 = 12$.

Transmitting Power

A belt or a chain going in a loop over two or more pulley drums makes all of them rotate when one is rotated. For example, a bicyclist rotates the special pulley drum, called a "crank," to which pedals are attached. The rotation of this crank is transmitted to the rotation of the rear wheel's crank, which makes the bicycle move. Using drums of different diameters, such as cranks on a sports bicycle drivetrain, can produce a force advantage.

Until the mid-twentieth century, factories typically used belts distributing power to individual machines from one central rotating drum, connected to a large steam, turbine, or animal-powered capstan engine. This power transmission system is called "line shaft." Because most industries have switched to compact electric motors, one is currently more likely to meet this type of a pulley in a museum or a history book. A human-powered capstan is also a popular science or historical fiction trope, used to demonstrate oppression, for example, in *Conan the Barbarian* and *Captain Blood*.

Further Reading

Boute, Raymond. "Simple Geometric Solutions to De l'Hospital's Pulley Problem." *College Mathematics Journal* 30, no. 4 (1999).

Hahn, Alexander. *Basic Calculus: From Archimedes to Newton to its Role in Science*. Emeryville, CA: Key College Publishing, 1998.

Rau, Dana. *Levers and Pulleys: Super Cool Science Experiments*. Ann Arbor, MI: Cherry Lake Publishing, 2009.

MARIA DROUJKOVA

Radio

Category: Communication and Computers.
Fields of Study: Algebra; Measurement; Representations.
Summary: Radio waves have numerous applications and are described, analyzed, encoded, and "jammed" using mathematics.

Radio is a means of sending information by transmitting signals using radio waves, which are a type of electromagnetic radiation with frequencies in the spectrum of approximately 3 kilohertz (kHz) or 1000 cycles per second, to 300 gigahertz (GHz), or 1 billion cycles per second. These units are named for German experimental physicist Heinrich Hertz. Radio waves are used not only to carry radio and television signals but are also used in many other common technologies including wireless computer networks, wildlife tracking systems, cordless and cellular phones, baby monitors, and garage door openers. One interesting way that mathematics connects to radio is through mathematically based radio shows, like *Math Medley*, which was hosted by Patricia Kenschaft. Mathematicians have also spoken on programs like National Public Radio's *Science Friday*.

Radio waves are sinusoidal, meaning that they are characterized by a smooth, repetitive oscillation whose function at time t can be described algebraically as

$$y(t) = (A)\sin(\omega t + \phi)$$

where A is the wave's amplitude (peak deviation), ω is the wave's angular frequency (described in radians per second), and φ is the wave's phase (where the wave cycle is at time $t = 0$).

Brief History and Unique Properties

In 1864, the British physicist James Clerk Maxwell predicted the existence of radio waves as part of his theory of electromagnetism. Hertz confirmed Maxwell's theory between 1886 and 1888 and is generally credited with being the first person to send and receive radio waves. Several individuals played an important role in developing a practical system of radio transmission including the Serbian-American engineer Nikola Tesla, who demonstrated wireless radio communication in 1893; the British physicist Oliver Lodge, who demonstrated the transmission of Morse Code using radio waves in 1894; and the Italian physicist Guglielmo Marconi, who in 1896 was granted the first patent for a radio. Radio communications between ships and coastal stations were in use by 1897, and the first radio time signal (used to synchronize clocks) was transmitted from a U.S. Naval Observatory clock in 1904.

Radio waves may be broadcast over long distances because of the Heaviside Layer (also called the "Kennelly–Heaviside layer"), a conducting layer in the ionosphere predicted independently in 1902 by the

The Radio Astronomy Explorer was a radio telecope placed in a moon orbit in 1973 to obtain radio measurements of the planets. (National Aeronautics and Space Administration)

British mathematician and physicist Oliver Heaviside and the British physicist Arthur Edwin Kennelly. The existence of the Heaviside Layer was established in 1924 by the British physicist Edward Appleton, who also determined that the height of this reflective layer was about 100 kilometers (62 miles) above the Earth's surface. The Heaviside Layer allows radio signals to follow the curvature of the Earth (rather than disappearing into space) because they are reflected by the Heaviside layer and thus "bounce back" to Earth.

Applications

Radio astronomy, which led to the discovery of objects such as pulsars and quasars, dates from the 1931 discovery by American physicist Karl Guthe Jansky of radio waves emitted from the Milky Way galaxy. American astronomer Grote Reber created the first radio frequency sky map in 1941, and in the 1950s, the British astronomers Martin Ryle and Antony Hewish produced two notable catalogues of celestial radio sources.

Historically, most radio broadcasts used one of two techniques for sending their signals: amplitude modulation (AM) or frequency modulation (FM). AM is the older technology (the first AM broadcast took place in 1906) and it was the dominant radio technology for most of the twentieth century. AM encodes information by modifying the amplitude of the transmitted signal. The technology for FM broadcasting, which encodes information by varying the frequency of the transmitted signal, was developed in the 1930s and became common by the late 1970s. The information in these analog signals is inherently part of the signal itself—the information influences the wave's shape, and thus information loss can occur with any disruption of the signal. One example is the audible static that occurs when a radio receiver begins to travel beyond the range of a radio transmitter. In the twenty-first century, digital modulation has been increasingly used to minimize this problem. Digital modulation transfers digitized information using a broad spectrum of radio frequencies—far more than the AM or FM systems. Further, each signal is sent many times, reducing the chance of interference and signal loss because separate bits from many streams may be pieced together. Further, since the radio waveforms are not altered by the information, multiple signals may be carried at the same time in the form of one composite signal that is decoded by the receiver, a technique called "multiplexing." Satellite radio systems take advantage of multiplexing and the wider angle of coverage to offer many hundreds of specialized channels across broad geographic areas. Television is also transitioning from analog to digital signals.

Radio transmissions are used for communication during wartime, but because a radio signal may be picked up by anyone with a receiver, various coding methods have been developed. One famous example is the code talkers used by the American Army during World War I and World War II. This program capitalized on the fact that Native-American languages such as Navajo and Choctaw were almost unknown outside those tribes and also developed a simple code for terms like "tank" and "submarine," which allowed them to code and encode messages rapidly and with little risk of comprehension by the enemy. Also in World War II, the German Army used mechanical circuits to encrypt information. Although supposedly unbreakable because of the large number of combinations possible, the British mathematician William Tutte was able to deduce the pattern of the encoding machines after British intelligence intercepted two long coded messages, each of which was transmitted twice (the second time with corrected punctuation).

Interference

Radio waves can be blocked by weather formations, geographic features, and many other natural phenomena. Further, if several stations are broadcasting on a similar frequency, they may interfere with each other. Use of an antenna tuned to a particular frequency (so it will pick up the signal at the frequency more strongly than signals at other frequencies) and aimed at the source of the signal can improve reception. Radio signals can be deliberately jammed by broadcasting noise on the same frequency as the signal. For example, the Soviet Union regularly jammed broadcasts by Radio Free Europe and Voice of America.

To minimize unintentional interference, different parts of the radio spectrum are reserved for different uses and broadcast stations are assigned specific frequencies for their use. In the United States, AM radio uses frequencies from 535 to 1700 kHz, and FM radio uses frequencies between 88 megahertz (mHz) and 108 mHz. A radio station that identifies itself as "90.7 FM" is broadcasting at the frequency of 90.7 mHz, or 90,700,000 cycles per second (technically, 90.7 mHz is the station's mean frequency). Other parts of the spectrum are reserved for other uses. For instance, 30–30.56 mHz is allocated for military air-to-ground and air-to-air communications systems for tactical and training operations and for land mobile radio communication in support of wildlife telemetry and natural resource management.

Further Reading

Regal, Brian. *Radio: The Life Story of a Technology.* Westport, CT: Greenwood Press, 2005.

Richards, John. *Radio Wave Propagation: An Introduction for the Non-Specialist.* New York: Springer, 2008.

Weightman, Gavin. *Signor Marconi's Magic Box: The Most Remarkable Invention of the 19th Century and the Amateur Inventor Whose Genius Sparked a Revolution.* Cambridge, MA: Da Capo Press, 2003.

Sarah Boslaugh

Robots

Category: Architecture and Engineering.
Fields of Study: Algebra; Data Analysis and Probability; Geometry; Number and Operations.
Summary: Robots, their motion driven by mathematical algorithms and coordinate or polar geometries, have long been incorporated into society and popular culture.

Robots and robotic systems are increasingly commonplace in many areas of daily life, such as manufacturing, medicine, exploration, security, personal assistance, and entertainment. In general, a robot is a mechanical device that can perform independent tasks guided by some sort of programming. Sometimes, robots are intended to replace humans in tedious or hazardous tasks. In others tasks, such as some surgeries, robots may actually exceed human capabilities. For many, the word "robot" brings to mind both futuristic androids, which are robots that are designed to look human and cyborgs, which contain both mechanical and biological components. Robots used in many industrial applications, such as in medicine, bomb disposal, and repetitive jobs, rarely resemble humans. However, several humanoid robots and robots that realistically mimic the look and behavior of animals have been produced.

In 2008, a Japanese play was written and produced for both robots and human actors, and robot animals have sometimes been marketed as replacements for biological pets. The word "robot" can also refer to software-like Web crawlers that run automated tasks over the Internet to gather data, though "bot" is a more common name. The field of robotics generates many interesting problems in both theoretical and applied mathematics and benefits from the contributions of mathematicians. For some, the ultimate quest in the twenty-first century and beyond is to develop materials, technology, and algorithms to create robots that meet or perhaps exceed human levels of perception, behavior, and intelligence. Nano-robots, which are ultra-small robots about the size of a nanometer, might one day be developed for tasks like hunting and destroying cancer cells.

Brief History

Playwright Karel Capek is typically credited with introducing the word "robot" from the Czech word for "laborer," in his 1920 play *R.U.R.* (Rossum's Universal Robots). Another writer who popularized robots was Isaac Asimov, who introduced the term "robotics" in his 1941 short story *Runaround*. However, robotic devices can be found much farther back in history. One early robotic device was a water clock produced by the Babylonians, which used the mathematics of volumes and rates of water flow to calculate time. Greek mathematician Hero of Alexandra described the use of weights and ropes to construct a mobile cart that could be programmed to move along a path. In the thirteenth century, Muslim mathematician and scientist Abu Al-'Iz Ibn Isma'il ibn Al-Razaz Al-Jazari created a set of programmable musicians. The drummer was operated by a rotating shaft that manipulated levers to produce rhythms. Around 1495, Italian painter and mathematician Leonardo da Vinci used his knowledge of the mathematics of anatomy and bodily movement to sketch designs for a warrior robot outfitted in medieval armor.

Interest in robotics accelerated in the nineteenth century as early computer technology with punch cards began to be incorporated into systems such as that used for the Jacquard loom, named for Joseph Jacquard. Others, such as Pafnuty Chebyshev, studied the theoretical mathematics of linkages, inventing the Chebyshev linkage that converts rotating motion to approximate straight-line motion. Charles Babbage's mathematical engines were some of the first mechanical computers.

These engines used finite differences to calculate the values of polynomials. Such inventions were forerunners of computer-controlled robot technology that quickly progressed in the mid-twentieth century to transistors and integrated circuits. Mathematician Norbert Weiner is often known as the "father of cybernetics," which is the science of self-regulating feedback systems, for his work and 1948 book *Cybernetics: Or Control and Communication in the Animal and Machine*.

Cybernetics is not synonymous with artificial intelligence or robotics, but this mathematical discipline is essential for environmentally responsive or adaptive robots. Some other areas of mathematics that have contributed to the development and implementation of robots included algebraic and differential geometry, which is used to help solve problems, such as orientation and movement in three dimensions; partial differential equations, which are used to model many aspects of behavior; optimization algorithms to help sequence tasks; combinatorics, which is used to investigate modular components and systems; and Bayesian statistical methods, named for Thomas Bayes, which can be employed in dynamic perception and machine learning.

Robotic Motion

In the twentieth and twenty-first centuries, many robots are complex, electromechanical devices that move and interact with physical objects, often replacing or augmenting human actions by carrying out certain tasks. Some mobile robots use articulated legs or wheels. Somewhat more common are stationary robotic arms with joints that allow for motion similar to the way joints allow human limbs to move. Having more joints increases the possible angles for movement and degrees of freedom, and hence increases fluid motion and accuracy. Articulated robots, used widely in various industries to perform tasks such as welding components or spray-painting parts, look much like human arms and have at least three joints. If the joints are slide-only, called "prismatic joints," then the robot arm can reach any position in a rectangular workspace by means of translations. If one joint is hinged, which is called a "revolute joint," then all points within a cylindrical workspace can be reached by a combination of rotation and translation. If two of the joints are hinged, a robot arm with a polar geometry is achieved. Inventor George Devol and engineer Joseph Engelberger developed one of the first modern-day programmable robots, Unimate, which began operation in 1961 at a General Motors plant. In 1969, Stanford University stu-

In November 2010, Robonaut 2 was brought to the International Space Station where it will remain as the first humanoid robot to work in space. (National Aeronautics and Space Administration)

dent Victor Scheinman created the predecessor for all robotic arms, the Stanford arm.

Mathematical programming and calibration for proper movement of robots depends on kinematics, which is the study of motion; and dynamics, which is the study of how force affects motion. With articulated or jointed robots, for example, the mathematics of kinematics is at the heart of positioning, collision avoidance, and redundancy. Direct kinematics makes use of given joint values to determine the end position that a robot arm may achieve. The mathematics of inverse kinematics is used to determine the required values for the joints when the end position of the robotic arm motion is known. Getting the robot arm to the right position is only half of the mathematical problem. The other half involves calculating forces using dynamics. For example, a robot designed to fight fires would need motors to move the robot and its arms. Calculations incorporated in determining which motors to use would involve dynamics. Inverse dynamics would help determine the required values of forces to generate the desired acceleration of the robot or its components. The movement involved in robotics most often occurs in three-dimensional space, so geometry plays a role in the positioning and movement of robots. Matrices can be used to represent the points through which robots navigate. These algebraic representations are then reviewed and coordinated using sophisticated applications of basic calculus principles, like differentiation, to ensure maximum efficiency when designing and operating robots.

Movement and action in robots are driven by algorithms. Some robots respond to direct human input from keyboard commands or from haptic devices that respond to tactile or body motion. Others autonomously perform programmed tasks. Some robots are "smart" or "intelligent," meaning that they are able to sense and adapt to their surroundings while completing their tasks. Even then, these robots are able to accomplish tasks only because they have been programmed to do so. For example, "smart" mobile robots make use of a variety of sensors with terrain-identification and obstacle-detection programs using input data and probabilistic models to guide trajectory and avoid collisions. Probabilistic robotics is increasingly of interest, with the goal of developing algorithms that facilitate accurate autonomous decision making in the face of real-work complexity and uncertainty, which would increase the reliability of automated behavior and more closely replicate the type of processing that occurs in the human brain.

Robots: Fiction and Fact

Robots are widely used in entertainment, especially science fiction. Mary Shelley's 1818 novel *Frankenstein* is cited by some as showing that scientific creations able to perform human tasks long preceded television and movies. Some well-known examples include C-3PO from the *Star Wars* series and Wall·E from the 2008 Pixar movie of the same name. Data, from the 1987–1994 television series *Star Trek: The Next Generation*, is an example of a fictional android. The Borg species from the *Star Trek* series and the Terminator robot from *The Terminator* movie series are examples of cyborg characters, usually hybrid humans whose biological capabilities are sustained or enhanced through robotic elements—though the Terminator may be thought of by some as a robot enhanced by biology. Enhancing human capabilities through robotic elements, like pacemakers and prosthetic devices, is common in the twenty-first century. However, the medical applications of robotics have not focused on humans achieving superhuman powers (as is done in fiction) but rather on helping those with medical conditions and disabilities.

Robots in Education

Robots are often used in schools to motivate learning of mathematics concepts, such as two- and three-dimensional coordinate geometry. The roBlocks construction system was developed by computational design scientists Mark Gross and Eric Schweikardt. Users can build robots using modular sensor, logic, and actuator blocks to study concepts like kinematics, feedback, and control. They can also create their own control programs to further explore robot mathematics and dynamics. The Lego Group produces a robotic construction and programming system called Mindstorms NXT that has been marketed for both education and entertainment.

Further Reading

Craig, John J. *Introduction to Robotics: Mechanics and Control*. 3rd ed. Upper Saddle River, NJ: Prentice Hall, 2004.

Murray, Richard M., Zexiang Li, and S. Shankar Sastry. *A Mathematical Introduction to Robotic Manipulation.* Boca Raton, FL: CRC Press, 1994.

Thrun, Sebastian, Wolfram Burgard, and Dieter Fox. *Probabilistic Robotics.* Cambridge, MA: MIT Press, 2005.

Deborah Moore-Russo
D. Keith Jones

Roller Coasters

Category: Games, Sport, and Recreation.
Fields of Study: Algebra; Calculus; Geometry; Measurement.
Summary: Roller coasters are mathematically designed to provide safe and thrilling rides.

Roller coasters are entertainment rides designed to put the rider through loops, turns, and falls, inducing sudden gravitational forces. The rapid ascents and descents coupled with sharp turns create momentary sensations of weightlessness. One known precursor of roller coasters are seventeenth-century Russian ice slides, which sent riders down a tall, ice-covered incline of roughly 50 degrees. Modern roller coasters can be traced to the late 1800s. As of 2010, Ohio's Cedar Point held the record for most roller coasters (17) in a single amusement park.

Conservation of Energy

The law of conservation of energy states that energy can neither be created nor destroyed, but can only be converted from one form to another. Roller coasters exploit this law by converting the potential energy gained by the car as it ascends to the top of a hill into kinetic energy as it descends and goes through the turns and loops. The potential energy of the car at the top of the loop is given by

$$E = m \times g \times h$$

where E is the total potential energy (joules), m is the total mass of the car (kg), g is the acceleration due to gravity (9.8 m/s^2), and h is the height (m).

For example, consider a roller coaster car weighing 2200 pounds perched at the top of Cedar Point's Top Thrill Dragster, which is about 426 feet high. The car, at this point, has accumulated $1000 \times 9.8 \times 130 = 1{,}274{,}000$ joules or 1.2 megajoules of energy—the same amount of energy released by the explosion of a quarter kilogram of TNT. This potential energy is converted into kinetic energy as the car hurtles down the loops.

As the car expends potential energy, it is converted into kinetic energy, propelling it forward. In an ideal situation where there is no friction or air drag, the car would travel forever. However, because of friction and other resistive forces, the car decelerates and finally stops when it has expended all its potential energy.

Centripetal Force

Centripetal force is responsible for keeping the rider glued to the seat as the car executes turns and loops and even puts the rider upside down. Centripetal and centrifugal forces act on a body that is traveling on a curved path. Whereas centrifugal force is directed outwards, toward the center of curvature, centripetal force acts inward on the body.

G-Force and Loop Design

G-forces are non-gravitational forces, and can be measured using an accelerometer. Humans have the ability to sustain a few *g*'s (a few times the force of gravity), but deleterious effects are a function of duration, amount, and location of the g-force. Many roller coasters accelerate briefly up to six *g*'s, depending on the shapes, angles, and inclines of loops, turns, and hills. Early roller coaster loops were circles. To overcome gravity, the cars entered the circle hard and fast, which pushed riders' heads continually into their chests as the coaster changed direction. In the 1970s, coaster engineer Werner Stengel worked with National Aeronautics and Space Administration (NASA) scientists to determine how much force riders could safely tolerate. As a result of this and other mathematical investigations, he began to use somewhat smoother clothoid loops, which are based on Euler spirals, named for Leonhard Euler. In 2010, using the same equations that describe how planets orbit the sun, mathematician Hanno Essén drew a new and unique series of potential rollercoaster loops. Riders would get the thrilling visual experience of a loop without any of the typical jolting and shaking, because the force that riders would feel pushing them into their seats would stay exactly the same all the way around the loop.

Further Reading

Alcom, S. *Theme Park Design: Behind The Scenes With An Engineer*. Orlando, FL: Theme Perks Press, 2010.

Koll, Hilary, Steve Mills, and Korey Kiepert. *Using Math to Design a Rollercoaster*. New York: Gareth Stevens Publishing, 2006.

Mason, Paul. *Roller Coaster!: Motion and Acceleration*. Chicago: Heinemann-Raintree, 2007.

Rutherford, Scott. *The American Rollercoaster*. Norwalk, CT: MBI, 2000.

Ashwin Mudigonda

Search Engines

Category: Communication and Computers.
Fields of Study: Algebra; Data Analysis and Probability; Number and Operations.
Summary: Using complex and sometimes proprietary algorithms, search engines locate and rank requested information, usually on the Internet or in a database.

Search engines are used for finding information from digitally stored data. Based on a search criterion like a word or phrase, search engines find information from the Internet and personal computers and present search results appropriately. A search engine is a very efficient tool for effortless finding information from millions of Web sites and their Webpages. For example, information on movies or weather forecast from the Internet can be easily found using search engines. To sort through vast amount of data, search engines use statistics, probability, mathematics, and data analysis.

Types of Search Engines

Different types of search engines are developed for different purposes. The simplest one is a desktop search engine, which is used for finding information stored within a computer. An enterprise search engine searches for digitally stored information within only one organization. A Web search engine looks for information on the World Wide Web (WWW). Sometimes, federated search engines are used for searching online databases or related items. Though there are different types, the term "search engine" generally refers to Web search engines.

Search Mechanisms

Searching for a word or phrase in a document in a computer is very simple and sophisticated search engines are not needed for this. A program simply reads the whole or selected part of the document, looks for where the intended word or phrase is located, and highlights the locations in the document.

Desktop search engines perform more complicated searches. These engines read all files and folders kept in the computer to collect information and index them. Indexing is a method of storing information about files and folders considering several factors like file names, contents, types, authors, and locations of files. It uses mathematical manipulations involving numbers, operations, and data mining. Once indexing is finished, the engine follows that index for searching. For example, if the word *algebra* is searched in a computer, the engine reads the index and tries to find out where the word *algebra* is located (if anywhere), and it shows the resulting files or folders.

The most complicated and interesting search engines are Web search engines. The Web contains billions of Web pages, and each page contains information. These search engines search for information from almost all of them. These engines generally work in three major steps: (1) collecting information from the Web, (2) indexing, and (3) presenting search results.

For reading Webpages and collecting information, almost all Web search engines have their own computer program, often called a "crawler." A Web search engine may have one or more crawlers. The information collected by crawlers contains subject matters, hyperlinks, images, and other information. Next, the search engines index the collected data and store them for future retrieval. The index is like a giant catalogue and involves huge mathematical applications to prepare. When a search criterion is given for searching, search engines follow this index; they find which Webpages contain the information and present results as lists of links to those pages.

A challenging task for Web search engines is to present the search results properly and quickly. While showing the results, it is expected that the more relevant pages corresponding to the search criterion should appear earlier than less relevant pages. Different search engines have different algorithms for arranging pages based on relevance. For example, the Google search engine uses an algorithm called PageRank for this purpose. It uses probability, data analysis, matrix algebra, and related fields.

Examples of Search Engines

Web search engines began to be developed in the 1990s and are constantly improving to handle the increasing size and content of the Web. Many of the individuals who develop and refine search engines have degrees in mathematics. Popular search engines like AltaVista (launched in 1995), Google (1998), Yahoo Search (2004), and Bing (2010) are only a few examples. Google Desktop, GNOME Storage, Windows Search, and Easyfind are among the most popular desktop search engines, while OpenSearchServer and DataparkSearch are good examples of enterprise search engines.

Further Reading

Levene, Mark. *An Introduction to Search Engines and Web Navigation.* London: Pearson, 2005.

Voorhees, E. M. *Natural Language Processing and Information Retrieval.* New York: Oxford University Press, 2000.

Sukantadev Bag

Segway

Category: Travel and Transportation.
Fields of Study: Algebra; Calculus; Geometry; Measurement; Number and Operations.
Summary: The Segway is a personal transporter built on the principle of dynamic equilibrium.

The Segway is an electric, two-wheeled personal transportation device that utilizes principles of balance and equilibrium both to create and control its motion. The Segway transporter was developed in part to combat the congestion and pollution caused by automobiles. In many cities, Segway tours are now alternatives to walking or bus tours. The Segway is often cited as an

Riders lean forward to make the wheels move forward and lean back to stop the Segway. (iStockphoto)

application of a classical dynamical systems problem: the inverted pendulum problem. It can also be referenced in illustrating the phenomenon of dynamic stability that also occurs in human walking.

Inverted Pendulum

In a traditional pendulum problem, the pendulum is composed of a mass attached to a string that is itself attached to a pivot point. In this case, the mass hangs below the pivot point. The position in which the mass hangs below the pivot point is stable—the pendulum eventually returns to that position even if pushed away from that position. In fact, it is relatively easy for the pendulum to rest in this equilibrium position. In an inverted pendulum problem, the situation in which the mass is above the pivot point is considered. Frequently, one can visualize this scenario as a "cart and pole." With the cart at rest, if the pole is perfectly positioned, it will stand upright on top of the cart. However, this condition is unstable; if the pole is moved away from this resting position, it falls.

An interesting property about the inverted pendulum (or cart-and-pole) problem is that as long as the base, or cart, is resting, the upright position is unstable. However, if the base or cart is in motion, oscillating at the right frequency, the upright position becomes stable. Imagine that the cart is moving forward and backward ever so slightly and very rapidly; in this case, the pole can remain upright. Now, the pole is in a dynamically stable position. This type of motion-induced stability is similar to what happens as humans walk. If an individual leans forward with his or her feet firmly planted on the ground, the individual will fall. However, if the feet are allowed to move, the individual will not fall but instead will move forward (or backward, depending on the direction of the lean). Allowing the feet to move has made the leaning position dynamically stable. With the feet moving, it is much harder for the individual to fall.

Dynamic Equilibrium

The Segway transporter operates on this principle of dynamic equilibrium. Riders lean forward to cause the wheels to move forward and lean back to cause the Segway to stop or reverse. The wheels and base are dynamically moving to keep the rider in an upright position instead of falling to the ground. Balance sensors in the base of the Segway regulate and control the motion by incorporating the pitch angle (or tilt) of the rider, the change in pitch angle, the wheel speed, and the wheel position. Mathematicians, physicists, and engineers relate all these variables through differential equations describing motion; these equations have long been studied in each of these fields. The Segway transporter is one example of a project resulting from the interplay of all three fields.

Further Reading

Kalmus, Henry P. "The Inverted Pendulum." *Journal of Physics* 38, no. 7 (1970).

Kemper, Steve. *Code Name Ginger: The Story Behind Segway and Dean Kamen's Quest to Invent a New World*. Cambridge, MA: Harvard Business School Publishing, 2003.

Tweney, Dylan. "Dec. 3, 2001: Segway Starts Rolling." *Wired* (December 3, 2009). http://www.wired.com/thisdayintech/2009/12/1203segway-unveiled.

Vasilash, Gary S. "Learning From Segway: Innovation in Action." *Automotive Design & Production* (January 2006).

Angela Gallegos

Servers

Category: Communication and Computers.
Fields of Study: Algebra; Communication; Number and Operations.
Summary: Servers help users connect to networks, including the Internet.

ARPANET, the first network of time-sharing computers, was connected in 1969. In subsequent decades, technology developments and the increasing benefits of distributed, shared access spurred network growth, ultimately resulting in the Internet and World Wide Web. Most local, national, and global networks rely on servers, which manage network resources for client computers that are connected to it. A server may be a physical computer, a program, or a combination of hardware and software. In some cases, a system is a dedicated server. In other cases, software servers operate on multipurpose systems. A distributed server is a scalable grouping in which several computers act as one entity and share the work. In general, a network server manages overall network traffic, while specialty servers handle other tasks. CERN httpd (or W3C httpd), which debuted in 1990, is considered to be the first Web server. It was developed by scientists Tim Berners-Lee, Ari Luotonen, and Henrik Frystyk Nielsen at the European Organization for Nuclear Research (CERN). Servers and clients use communication protocols to exchange information to carry out tasks. There are server-to-server and client-server variations. Mathematicians, computer scientists, and others work to create technology and algorithms that make servers possible and increase their efficiency. They also study the properties of networks and servers, which facilitates advances in both mathematics and computers. For example, in a system with multiple parallel servers, jobs may be assigned to any server. Often, jobs are modeled with an exponentially distributed processing time or some other probabilistic distribution with some resource cost per unit of time. Mathematical methods may be used to find the optimal strategy for allocating jobs to servers to minimize costs.

Function

The term "server" does not describe a specific type of computer in the same sense that "desktop" or "Windows machine" does. When used in reference to hardware, a server is any computer running a server program, which can—and in practice does—include all configurations and operating systems. Since the 1990s and the increased demand for Internet services, there have been more and more computers that have been designed specifically to be used as Internet servers. Because they need to run for long periods of time without interruption, they must be durable, reliable, and have uninterruptible power supplies. Typically, hardware redundancy is incorporated, so that if a hard drive fails, another one is automatically put on line—a feature rarely found in personal computers. There is also a great deal of server-specific hardware, such as water cooling systems, which help reduce heat, and Error-Correcting Code (ECC) memory, which corrects memory errors as they happen, preventing data corruption. Many components are designed to be hot-swappable, meaning that they can be replaced while the server runs—without needing to power it down. Furthermore, ordinary server operations including turning the power on or off can often be conducted remotely; for example, from a home computer. Some system operators maintain watch over multiple servers in multiple locations and physically visit the site only when necessary because of a crisis.

Communication

Sockets are the primary means by which network computers in a network communicate. They are the endpoints of the flow of interprocess communication (IPC) and provide application services. They are also the place where many security breaches take place. Mathematicians and computer scientists study the different socket types and their states to understand how they work and to improve function and security. Servers create sockets on start-up that are in listening state, waiting for contact to be made by client programs. For instance, a Web browser, like Firefox, is a client program used to access content from Web servers. Most servers connected to the Internet use a protocol known as Transmission Control Protocol (TCP), developed by computer scientists Vinton Cerf and Robert Kahn for ARPANET. An Internet socket is referred to by its socket number, a unique integer that includes Internet Protocol (IP) address and socket number. Listening sockets using TCP are usually assigned the remote address 0.0.0.0 and the remote port number 0. TCP servers can serve multiple concurrent clients by creating what is called a "child process" associated with each client and establishing TCP connections between child processes and clients. Each connection uses a unique dedicated socket. Two communicating sockets—the local socket created by the server and the remote socket of the client—are called a "socket pair," and their activity is referred to as a "TCP session."

A common feature of Web servers is server-side scripting, which allows Web pages to be created in response to client activity. For instance, a search for a book on Amazon.com results in a unique search results page. Without this capacity, every possible search would need to be conducted in anticipation of client needs.

Further Reading

Chevance, Rene. *Server Architectures: Multiprocessors, Clusters, Parallel Systems, Web Servers, and Storage Solutions*. Oxford, England: Elsevier, 2005.

Dshalalow, Jewgeni H. *Frontiers in Queueing Models and Applications in Science and Engineering*. Boca Raton, FL: CRC Press, 1997.

Gray, Neil A. B. *Web Server Programming*. Hoboken, NJ: Wiley, 2003.

Bill Kte'pi

Shipping

Category: Business, Economics, and Marketing.
Fields of Study: Geometry; Measurement; Number and Operations.
Summary: A variety of mathematical concepts, including packing, routing, and tracking, are necessary to make the process of shipping goods more efficient.

The shipping and delivery industry is a vast global business that is responsible for delivering packages, postal mail, and commercial cargo all over the world. In 2009 alone, express delivery companies made $130 billion in revenue worldwide, the U.S. Postal Service delivered 177 billion pieces of mail, and ocean liners transported more than $4.6 trillion worth of goods between nations. With so many items being delivered to so many different places, there is a need for mathematics to help manage the complex delivery network and ensure that deliveries are made correctly, safely, cheaply, and quickly. Mathematics has had a significant impact in three key areas of the shipping industry: container packing, vehicle routing, and package tracking.

Container Packing

To minimize transportation costs and maximize profit, a shipper would naturally prefer to pack cargo into as few shipping containers as possible. Determining the optimal way to arrange items in a container is a deceptively difficult problem. Given a set of differently sized objects, the Bin-Packing Problem is to find the order in which to place the objects so that they fill the minimum number of bins. Testing every permutation of packing the objects would be too time-consuming, so an efficient and simple algorithm is required.

A common packing procedure is the First-fit algorithm, where the objects are ordered from largest to smallest, and each object is placed in the first available bin that will hold it. It can be proven mathematically that this algorithm is not guaranteed to produce the optimal packing. In the worst case, the result can be far from optimal and require the use of more bins than a more sophisticated packing. The First-fit algorithm is an example of an approximation algorithm, which means it produces a good approximate answer but not necessarily the optimal arrangement of objects. Other more sophisticated bin-packing algorithms have been developed, but as of 2010, no efficient algorithm was known that always produced the optimal packing.

In practice, there are more considerations to packing shipping containers. Some packages will be irregularly shaped and do not stack well. Some cargo is fragile and must be secured separately. Sometimes, a delivery vehicle will make several stops, so the packages that are delivered first should be packed into a container last to make them easily accessible.

Through World War II, most cargo was shipped in wooden crates of various sizes. A big step forward came in 1956, when trucker Malcolm McLean patented the modern shipping container made of corrugated steel. This sturdy container was easier to move between truck, rail, and ocean liner. More importantly, having a standard-size container meant that packing procedures could be standardized. Prior to 1956, it was estimated that loose cargo cost $5.86 per ton to load. After the standardized container was introduced, it was estimated the loading cost dropped to 16 cents per ton, a 3600% improvement.

Vehicle Routing

Cargo travels by a variety of transportation modes, including truck, rail, air freight, and ocean liners. The goal of routing is to determine a vehicle for each piece of cargo to be delivered and then find the shortest delivery route for each of the vehicles. The Traveling Salesman Problem is a simple mathematical example of a routing problem. In practice, the value of a route is not determined by just the distance. The problem is complicated by considerations such as personnel, fuel costs, traffic, tolls, and tariffs.

NP-Complete Problems

The NP-complete set is a list of mathematical problems for which there is no known fast algorithm for solving the problem exactly. The Traveling Salesman Problem is an example of a NP-complete problem. While there are fast algorithms for finding a good answer, the only known algorithm for finding the single shortest route is extremely slow. However, just because there is no known fast algorithm for solving these problems, it does not mean that such an algorithm does not exist. In 2000, the Clay Mathematics Institute offered a $1 million prize to anyone who could devise an algorithm that would solve an NP-complete problem quickly or prove that no such algorithm exists. While not technically an NP-complete problem, the Bin-Packing Problem is in a related category of problems known as "NP-hard."

Mathematical analysis of delivery routes can lead to huge improvements in shipping efficiency. As the first Postmaster General of the United States, Benjamin Franklin ordered careful surveying of delivery routes, refined the post office accounting practices, and increased public access to mail. Under this new system, the U.S. Postal Service became profitable for the first time, and it is estimated that the mail delivery time between major cities was cut in half.

The routing problem is an example of a problem studied in operations research, the branch of mathematics that studies the cost-effectiveness of decisions made by corporate management such as scheduling and personnel assignments. The field of operations research has its origins in World War II, when the Allied Forces were interested in coordinating the manufacturing and organization needed to mobilize the military. One of the early researchers in operations research was Tjalling Koopmans, who proposed a mathematical model for the routing problem for shippers.

Package Tracking

It is important for a shipper to carefully track a package until it reaches its destination. A common system for identifying a package is the barcode. By encoding the destination as a sequence of black and white bars, the packages can be sorted quickly by automated sorting machines equipped with laser scanners. The U.S. Postal Service has developed a special barcode that encodes the address as a sequence of short and tall black bars. The mail is first read by an Optical Character Recognition (OCR) program, which translates the handwritten address into a barcode. The barcode is stamped onto the package and then automatically sorted to be sent to the next distribution center.

Radio-frequency identification (RFID) is a tracking technology that could potentially have a large impact on the shipping industry. A small electronic tag that emits a radio signal would be placed on each item to be shipped. Generally, this tag is a microchip just a few millimeters on a side. Potentially, this microchip would allow a shipper to determine the entire contents of a shipping container without ever opening the container. However, the technology still needs to be refined to make RFID a cheaper alternative to the barcode. Furthermore, since an item could theoretically still be tracked after the delivery is made, RFID technology is somewhat controversial because of privacy concerns.

> ### Tjalling Koopmans (1910–1985)
>
> Tjalling Koopmans was a Dutch economist who helped develop the mathematical field of operations research. Working for the British Merchant Shipping Mission in the 1940s, Koopmans derived a mathematical model for finding the most cost-effective shipping routes. Later, he became a professor of economics at University of Chicago and then at Yale University. In 1975, Koopmans received the Nobel Prize for Economics for developing mathematical tools for the analysis of corporate management and efficiency.

Further Reading

Hillier, Frederick. *Introduction to Operations Research*. New York: McGraw-Hill, 2009.

Lodi, Andrea, Silvano Martello, and Daniele Vigo. "Recent Advances on Two-Dimensional Bin Packing Problems." *Discrete Applied Mathematics* 123 (2002).

Palmer, Roger. *The Bar Code Book*. Peterborough, NH: Helmers, 2007.

Roberti, Mark. "The History of RFID Technology." *RFID Journal* (2005).

Todd Wittman

Skyscrapers

Category: Architecture and Engineering.
Fields of Study: Algebra; Geometry; Measurement.
Summary: Mathematicians and engineers work together to design and build skyscrapers.

A skyscraper is a building noteworthy for its great height. As the name suggests, the building appears to touch the sky. There is no agreed-upon minimum height that classifies a building as a "skyscraper"; the term is used for any building that commands attention because of its height. Many people are fascinated by building, visiting, and measuring skyscrapers. The

Eiffel Tower, designed by engineer Gustave Eiffel, revolutionized civil engineering and architectural design. In the design of a skyscraper, architects and engineers must consider load distribution and the impact of the wind and earthquakes. Scientists and mathematicians also investigate how to improve features such as seismic dampers. Many skyscrapers resemble rectangles or pyramids, but they may have other geometries, like the plan for the Helicoidal Skyscraper in New York or the sail-shaped skyscraper in Dubai—the Burj al-Arab Hotel. In Tokyo, St. Mary's Cathedral incorporates eight hyperbolic parabolas, and the HSB Turning Torso in Sweden uses five-story cubes that twist as they rise, with the top cube 90 degrees from the bottom cube. Buckminster Fuller proposed a city consisting of huge floating spheres, which he called Cloud Nine. The Wing Tower in Scotland was designed to rotate at the base in order to respond to changes in the direction of the wind. Proposed dynamic skyscrapers allow each floor to rotate independently, creating changing shapes, and using turbines to harness the power of the wind. There are various ways of ranking skyscrapers by height, and these buildings have other characteristics that can be quantified as well. Mathematician Shizuo Kakutani invented a mathematical skyscraper in ergodic theory called a "Kakutani skyscraper," so named because the mathematical process resembles the floors of a skyscraper. Students in some mathematics classrooms play a multiplication skyscraper game.

History

Throughout history, there have been buildings that were considered unusually tall, including pyramids, towers, and religious structures. The 10-story Home Insurance Building in Chicago, designed by William Le Baron Jenney and completed in 1885, is considered by many to be the world's first skyscraper. A variety of technological developments made the first skyscrapers possible. These included the mass production of steel, the invention of the elevator, the ability to achieve water pressure at altitude, the fireproofing of flooring and walls, and the development of reinforced concrete. The 792-foot Woolworth Building in New York City, completed in 1913, was typical of how skyscrapers would be constructed for the rest of the twentieth century. It had a steel skeleton and a foundation of concrete. Modern skyscrapers typically have a frame that supports the building's weight, with walls suspended from the frame.

This feature distinguishes them from smaller buildings in which the walls are usually weight-bearing.

The Empire State Building in New York City reigned for 41 years as the world's tallest skyscraper and entered the public consciousness when the 1933 film *King Kong* depicted a giant ape that climbed the building. The movie had innovative special effects, including the use of scale modeling. In the twenty-first century, numerous television and FM radio stations transmit their signals from atop the Empire State Building and from skyscrapers in other cities.

Measurement

There are many different ways to measure the height of a skyscraper. It can be measured by the number of floors, highest occupied floor, spire height, or total height including such things as an antenna. Consequently, different figures can be found for the height of a single skyscraper. When lists of the world's tallest skyscrapers are published, a single skyscraper often ranks in different places on lists that use different rules of measurement. For example the Willis Tower in Chicago, formerly known as the Sears Tower, is the world's second tallest building when ranked by number of floors or when antennae are included, but it places seventh worldwide when spires are counted, but antennae are not.

Since 1998, a number of skyscrapers in Asia have surpassed the tallest American buildings in height. The Burj Khalifa, which opened in 2010 in Dubai, United Arab Emirates, is the world's tallest skyscraper as of 2010, whether ranked by its 163 floors, its 2,093-foot highest floor, or its spire height of 2,717 feet. The progression of record skyscraper heights over time can be graphed and modeled by a regression equation.

Other Aspects

Skyscrapers are noteworthy for other quantities besides their heights. When known geometric solids are used to model a skyscraper's shape, the building's surface area can be estimated. Because of differences in elevation, a skyscraper often experiences measurably different weather conditions at its top and bottom. In addition to its noteworthy height measurements, the Burj Khalifa contains over 20 acres of glass, has over 5 million square feet of floor space, and has elevators that travel over 26 miles per hour. Tall buildings are known to sway slightly in windy conditions. A rule of thumb for estimating a building's sway is to divide its height by 500 to arrive at the amount of horizontal sway near the top of the building. In many skyscrapers, steel tubes, or bundles of tubes, give the building strength against this swaying. The distance one can see from the top of a skyscraper can be computed. When the curvature of Earth is considered, the sight line is tangent to Earth's surface. On a clear day it is possible to see over 100 miles from atop the world's highest skyscrapers.

Further Reading

Koll, Hillary, Steve Mills, and William Baker. *Using Math to Build a Skyscraper*. Milwaukee, WI: Gareth Stevens Publishing, 2007.

Wells, Matthew. *Skyscrapers: Structure and Design*. New Haven, CT: Yale University Press, 2005.

David I. Kennedy

SMART Board

Category: Communication and Computers.
Fields of Study: Geometry; Representations.
Summary: Interactive whiteboards use a touch-sensitive display to mimic the functionality of a whiteboard while enhancing the user's options.

The SMART Board is a brand of interactive whiteboard. Unlike traditional whiteboards and chalkboards, the SMART board does not require markers, chalk, or erasers. Instead, the SMART Board utilizes a projector and a touch sensitive display. The projector displays computer images onto the screen. The screen itself allows the user to interact directly with applications similar to a large touch screen. For instance, touching the screen is equivalent to left-clicking with a mouse. Typically, SMART Boards come with four digital pens and a digital eraser. These digital devices allow the user to write on the screen using digital ink. SMART Board interfaces are available for Windows and for the Mac operating system.

Development

David Martin and Nancy Knowlton, inventors of the SMART Board, initially devised the idea in 1986 and began promoting it in 1991. Knowlton previously taught accounting and computer science, while Martin has a bachelor's degree in applied mathematics and began his career working on computer simulations. The SMART Board was the first interactive whiteboard that gave users touch control of computer applications. In 2003, their company developed and later patented Digital Vision Touch (DViT) technology, which relies on concepts of three-dimensional geometry, such as projection, reflections, and parallel lines to effectively display information and allow the user to interact with the board. It uses digital cameras and sophisticated recognition algorithms to determine the position of the user's fingertip and to make a distinction between single clicking, double clicking, and drag and drop. These recognition algorithms differentiate it from other touch technologies, like tablet personal computers. As of 2010, SMART Board was the most popular interactive whiteboard on the market in the United States.

Advantages

There are several advantages to SMART Board technology in the mathematics classroom. First, lectures done using the SMART Board can be saved, which allows instructors to access information written minutes, weeks, or even years earlier. By exporting these files as a pdf or a similar universal format, the instructor can post classroom notes on their course Web page, allowing students to review notes from previous classes, either to prepare for a test or to catch up on material that was covered when they were absent. In addition, the digital images saved by the SMART Board can more easily be read and transcribed for students with disabilities. Further, images on the SMART Board can be individually selected and copied to additional pages, which allows complex mathematical formulas and diagrams to be reproduced accurately and quickly. SMART board systems are typically connected to computers, meaning any application that is accessible on the computer is available on the SMART Board. Instructors may access spreadsheets, word processors, and the Internet. For these reasons, SMART Boards can greatly enhance the educational experience for both the instructor and the student. SMART Board–type lectures can also be accomplished using a tablet computer installed with the appropriate software and a projector system.

Since its introduction in 1991, SMART Boards have been incorporated into classrooms of all levels from kindergarten to college. In addition, many corporate boardrooms feature SMART Boards allowing for interactive presentations. As of 2010, over 1 million SMART Board systems have been installed across the world. It is likely that SMART Boards and similar systems will continue to replace or supplement the more traditional whiteboards and chalkboards found in current classrooms.

Further Reading
Bitter, Gary G. *Using Computer Technology in the Classroom.* Boston: Allyn and Bacon, 1999.
Ellwood, Heather. "Practice Makes Perfect: Building Creative Thinking Skills in High School Mathematics." *EdCompass Newsletter* (March 2009).
Gooden, Andrea R. *Computers in the Classroom: How Teachers and Students Are Using Technology to Transform Learning.* San Francisco, CA: Apple Press, 1996.
McAndrews, Alyson. "Improving STEM Engagement. New Program Uses SMART Products to Help Students Collaborate and Connect." *EdCompass Newsletter* (December 2009).
Morrison, Gary R. *Integrating Computer Technology Into the Classroom.* Upper Saddle River, NJ: Merrill, 1999.
Peters, Laurence. *Global Learning: Using Technology to Bring the World to Your Students.* Eugene, OR: International Society for Technology in Education, 2009.
Price, Amber. "Ten Ways to Get Smart With SMARTboard." *Tech & Learning* (August 2008).

Robert A. Beeler

Smart Cars

Category: Travel and Transportation.
Fields of Study: Algebra; Geometry; Measurement.
Summary: A smart car is able to respond to the conditions it detects, such as sounding an alarm if it detects that a driver is becoming drowsy.

A smart car is also sometimes referred to as a "biometric car." The overall design and technology of such vehicles should incorporate many functions: protection of the driver and passengers, reliable and easy navigation, and better mechanical and fuel efficiency. Mathematicians, engineers, and many others are involved in the development of improved vehicle technology, including aerodynamics and computerized systems that use mathematical techniques from geometry; mathematical and computer modeling; and statistical analyses of data regarding safety, ergonomics, and consumer preferences. Methods from artificial intelligence, such as cellular automata, are also very useful. According to mathematician John von Neumann, cellular automata can be thought of as "cells" or agents that behave according to relatively simple sets of mathematical rules or algorithms. These rules include responses to neighboring cells' behaviors, making them useful in modeling many biological processes, like flocking birds or traffic.

Ideal Functions
In many peoples' minds, the primary purpose of a smart car should be to help a driver in ways that prevent accidents and encourage safe driving. For example, many car accidents occur because drivers do not realize that they are drowsy, so they consequently fall asleep at the wheel. A biometric smart car could alert drivers to such conditions by measuring eye movements relative to typical alert driver behavior to detect inattention and lack of scanning of the instruments and the road. Drivers that deviated too far from established safety norms would then be alerted. Other systems may involve a steering detector that responds to angular movements of the steering wheel that exceed a specified degree or a system that measures the angles of a driver's head and sound an alert if the head nods too far forward. In 2010, a Japanese company launched a system designed for commercial truck drivers that analyzes a driver's unique patterns and variability taking into account variables such as time. It then uses mathematical algorithms to proactively recommend rest breaks and measures to increase alertness and safety.

But What Actually Makes a Car Smart?
In addition to reactive systems like driver alertness warnings, some feel that a truly smart car should anticipate conditions to be avoided. Speeding when road conditions are poor or attempting to pass another car in low visibility could be predicted and avoided. Smart car systems would not only anticipate but also correct any anomaly so that a driver has time to recover. Further, they might suggest actions to a driver in advance of adverse conditions by monitoring the road and

weather. Aspects of these features are present in many models of cars at the start of the twenty-first century facilitated by the introduction of real-time technology, such as interactive maps and global positioning systems (GPS), which depend on external communication with the environment to provide data beyond the drivers' senses. For example, many agencies provide data on road grade and surface, work zones, hazards, or speed restrictions. A smart car also monitors its internal state, taking measures of aspects like tire pressure and fluid levels using electronic sensors—functions that used to have to be performed by hand.

Advanced instrumentation, once found mostly in luxury cars, is becoming commonplace in vehicles. These systems may include smart starting that relies on electronics embedded in the car's keys; biometric features, like fingerprint scans; or keyless entry that may also require a computer chip, code, or fingerprint to activate. Many hybrid gas–electric vehicles balance energy usage to obtain maximum performance in mileage. Future smart cars may automatically sense variables like weight distribution and suggest load adjustments for better balance and braking. There are even notions that future smart cars will be able to dynamically reshape their surfaces for maximum aerodynamic efficiency. There is work being done on systems such as neural networks that may monitor and analyze all driver decisions in order to better provide feedback for safety and performance for particular geographic regions. Networks within smart cars may also interact with other cars and "smart roads," which could use computer technologies and mathematical modeling or algorithms, coupled with control and communications features, to improve issues like road safety and traffic capacity by directing traffic and helping drivers make better and safer decisions.

Further Reading

Scientific American Frontiers. "Inventing the Future Teaching Guide: Smart Car." http://www.pbs.org/safarchive/4_class/45_pguides/pguide_701/4571_smartcar.html.

Volti, Rudi. *Cars and Culture*. Baltimore, MD: Johns Hopkins University Press, 2006.

Whelan, Richard. *Smart Highways, Smart Cars*. Norwood, MA: Artech House, 1995.

Julian Palmore

Social Networks

Category: Friendship, Romance, and Religion.
Fields of Study: Geometry; Number and Operation; Representations.
Summary: Social networks can be described and analyzed using graph theory.

A social network is a set of actors and the relationships that connect them. The actors are usually people, but may be other individual or collective actors, such as organizations, gangs, clubs, municipalities, nations, or social animals. Social network analysis is a cross-disciplinary method for analyzing social networks that integrates techniques from science, social science, mathematics, computer science, communication, and business. In keeping with its diverse origins, various types of social relationships have been studied using social network analysis, such as friendship, sexual relationships, kinship and genealogy, competitions, collaboration, and disease spread.

Sociogram, Sociomatrix, Graph, and Network

Modern social network analysis can be traced to Austro-American psychiatrist Jacob Levy Moreno, though many of the methods he employed in his work had been used before in a more piecemeal fashion. For example, French probabilist Irénée-Jules Bienaymé modeled the disappearance of closed families (for example, aristocrats) and family names in the nineteenth century. In his 1934 book *Who Shall Survive*, Moreno used diagrams he called "sociograms" to analyze friendships among girls in a training school in New York State. The girls were represented by points, and pairs of girls who were friends were connected by a line. In sociograms of relationships such as liking, which are not necessarily reciprocated, an arrowhead indicates the direction.

While very simple social networks can be analyzed by visual inspection, the power of social network analysis arises from the conceptualization of the sociogram as a mathematical graph, which can be analyzed using the concepts and methods of graph theory. Moreover, a graph can be represented by a square adjacency matrix in which each row and column represent a point, and the cell entries represent the presence or absence of lines between points. A graph can be generalized in several ways. Lines can have numerical values representing, for example, the strength, intensity, or frequency of a relationship. There can be multiple types of lines between pairs of actors, each representing one type of relationship. Actors can have various attributes with numerical values or qualitative labels. In social network analysis, real-life social networks are modeled by mathematical networks, then the properties of the networks are analyzed mathematically in order to draw conclusions about the structure of the social relationships.

Social Cohesion

Social cohesion is a fundamental issue in the social sciences; it is the "glue" or bond that holds a social group together. According to social network analysis, it is the network of social ties among members of the group. Therefore, to measure the level of social cohesion in a social group or subgroup, one must measure the extent of ties among the members. The density of ties among members is the simplest measure of connectedness. It is defined as the ratio of the number of actual ties to the number of possible ties and ranges from 0 to 1. In a network with one symmetric (undirected) type of tie, and k members, the total possible number of ties is

$$\frac{k(k-1)}{2}.$$

A network in which every actor is connected is called a "complete" graph, or a "clique."

It is easy to imagine four people all being friends with one another but less realistic to postulate a clique with a large number of members. For example, in a clique with 30 members, each would have to maintain ties with the 29 other members—an onerous task. Limits on human beings' time, energy, and memory constrain the number of people with whom they can maintain social ties. Therefore, social networks tend to become more sparse (the ties become less dense) as they become larger. Residents of a small village may know all the other residents, but this is impossible for city-dwellers. Thus, the village will tend to be more socially cohesive than the city. Density of ties has also been used to study social cohesion in such areas of social life as marriage, the family, small groups in laboratories, community elites, intercorporate relationships such as share ownership and interlocking directorates, scientific communities, and the spread of ideas and diseases.

The overall density of ties is a rather crude measure of connectivity and cohesion in a social network, because it is insensitive to local variations. Real-life social networks tend to contain islands of actors tied relatively densely to one another but disconnected or only loosely connected by sparse ties to other such islands. In the friendship network of a high school, there are likely to be a number of small cliques, perhaps loosely connected into larger subgroups that are in turn perhaps totally disconnected from one another. Detection of relatively cohesive subgroups in a network and delineation of their articulation into larger, less cohesive groups are a major theme in social network analysis.

Centrality

The centrality of an actor in a network is an important attribute, because centrality is associated with power, prestige, prominence, and popularity. In a network of ties representing flows or potential flows of valued social goods, such as information, a central actor is in a privileged position for both reception and transmission. The centrality of an actor may be intuitively evident from visual inspection of the drawing of a graph, especially if the graph is small or highly centralized. In larger graphs, a precise definition and formula are needed. The four main definitions of centrality are degree, closeness, betweenness, and power (or "eigenvector") centrality.

Degree centrality is the proportion of the other actors to which an actor is directly connected. The closeness centrality of an actor is based on how close the actor is to each of the other actors in the network and is the inverse of distance. The betweenness centrality of an actor is the extent to which the actor is "between" other actors; in other words, how often the shortest paths between pairs of other actors pass through the actor. Power centrality is defined recursively taking into account the power centrality of the actors to which an actor is adjacent.

Applications of Social Networks

The popular party game Six Degrees of Kevin Bacon tries to connect any movie actor to actor Kevin Bacon via costars in movies using the shortest number of steps. That value is an actor's Bacon Number. The Web site "The Oracle of Bacon," originally implemented in 1996, can be used to find the shortest path for any actor that can be linked to Kevin Bacon. The average path length as of September, 2010, was about three. It also allows a user to find a measure of centrality for the Hollywood network based around any actor in the database in terms of the average path length.

On a more personal level, the social network Web site Facebook includes an application called Friend Wheel that lets users visualize the interconnections among their friends as nodes and ties. Further, it selectively arranges the friends' names around the circumference of the wheel so that closely-knit groups or cliques are placed together and color-coded. Thomas Fletcher, a computer science and mathematics student at Bath University, developed the application and made it available in 2007.

Harkening back to Moreno's study, in 1995 a team of sociologists was the first to map the romantic and sexual relationships of an entire high school. Unlike similar adult networks, which tend to have several highly interconnected cores with loose interconnections (like airline hubs), the students were connected via long chains, more like a rural phone network. One chain linked 288 of the 573 romantically active students, though there were also many unconnected dyads or triads. Researchers attributed this finding in part to the often-elaborate teenage social rules about who may date. The surprising finding had important implications for educational practices like sex education programs.

Further Reading

Bearman, P. S., J. Moody, and K. Stovel. "Chains of Affection: The Structure of Adolescent Romantic and Sexual Networks." *American Journal of Sociology* 110, no. 1 (2004).

Furht, Borko. *Handbook of Social Network Technologies and Applications*. New York: Springer, 2010.

Moreno, Jacob L. *Who Shall Survive?* Washington, DC: Nervous and Mental Disease Publishing, 1934.

Wasserman, Stanley, and Katherine Faust. *Social Network Analysis*. New York: Cambridge University Press, 1994.

Peter J. Carrington
Sarah J. Greenwald
Jill E. Thomley

Spaceships

Category: Travel and Transportation.
Fields of Study: Algebra; Geometry; Measurement.
Summary: Every task involving spaceships, from their design to their launch to effective collision avoidance and communication, is mathematically intensive.

Spaceships, also called "spacecraft," are manned or automatic vehicles for flying beyond planet atmospheres. Different types of spaceships serve different purposes, including scientific or applied observations and data collection, exploration of celestial bodies, communication, and recreation.

According to the routes they take, spaceships can be classified as suborbital, orbital, interplanetary, and interstellar. According to the type of propulsion used, spacecraft engines can be designated as reaction engines, including rockets; electromagnetic, such as ion thrusters; and engines using fields, such as solar sails or gravitational slingshots. Mathematics is fundamental for spaceship design, operation, and evaluation. For example, mathematics is used to plan efficient trajectories, avoid collisions, communicate with satellites, transmit data over vast interplanetary distances, and solve complex problems like those that occurred in the famous Apollo 13 mission.

Mathematics in Spaceship Systems

Propulsion of a spaceship poses scientific and engineering problems that involve balancing forces and computing sufficient fuel, energy, work, and fluid mechanics. For any type of engine, the impulse it gives to the craft has to be calculated and compared to the craft's tasks such as leaving the gravity well of a planet or maintaining an orbit. For example, calculations for rocket engines involve variables including the changing mass of the craft as its fuel is spent, the efficiency of the engine, and the velocity of the rocket's exhaust. Solar sail theories involve such variables as radiation pressure of the light, the area of the sail, and the weight of the craft.

Mechanics and material sciences problems involved in the structure of spacecraft include withstanding the forces, temperatures, and electromagnetic fields involved in moving through space. For example, moving through a planetary atmosphere at speeds necessary to leave the planet's gravity well involves high temperatures from friction.

The guidance and navigation systems of a spaceship collect data and then compute position, speed, and the necessary velocity and acceleration to reach the destination. These systems also determine the relative position of the spaceship to nearby celestial bodies, which influence the craft's motion by their gravitational and electromagnetic fields. For example, mathematical description of a craft orbiting a planet includes the six Keplerian elements (for example, inclination and eccentricity) defining the shape, the size, and the orientation of the orbit, named for Johannes Kepler.

Most twenty-first-century spacecraft do not carry living organisms, but when they do, life support systems are necessary. Life support systems protect people, animals, or plants in the spaceship from harmful environments and provide air, water, and food. The design of life support systems involves biology, physiology, medical sciences, plant sciences, ecology, and bioengineering. Mathematical models for life support typically include calculations of safety margins, such as maximum allowable radiation doses. All organisms need some inputs (such as food, water, or oxygen) and produce some outputs depending on a variety of variables, such as activity levels. Spaceship ecosystem designers strive to produce waste-free, closed systems where water is reclaimed and plants are used to purify the air. Because of the complexity of the closed ecosystem problem, most current flights employ simpler, machine-driven life support systems.

Atmospheric Flight

Flight within an atmosphere presents very different problems compared to flight in a vacuum. The problems solved by applied mathematicians who study atmospheric flight include friction, turbulence, wing lift, aerodynamic shapes, and control of temperature. Spaceships launching or landing on planets have to be equipped for atmospheric flight. Because of differences in the vacuum and atmosphere flight requirements, many spaceships are designed to change their configuration when they cross atmospheric boundaries. For example, mathematical theories originally developed for origami are used to fold and unfold solar batteries, which can be used only in a vacuum because of their large area.

Escaping a Planet's Gravity

The problem of escaping the gravitational field of a large celestial body, such as Earth, is different from the problem of flight in space far from large bodies. For example, a certain velocity, called escape velocity, is required to leave any given planet. At the sea level of Earth, the escape velocity is about 11 kilometers per second (km/s) or 7 miles per second (mi/s). However, spaceships usually fly slower at first. The escape velocity is inversely proportional to the square root of the distance from the planet's center of gravity. Spaceships leaving the Earth reach these lower escape velocity levels at some distance from the surface. For comparison, the escape velocity from the Sun is about 600 km/s (373 mi/s) and the speed record as of 2010 for a spacecraft leaving the Earth is about 16 km/s (10 mi/s). This means that flights near the Sun are not technologically possible in the early twenty-first century. The escape velocity of a black hole is greater than the speed of light (over 300,000 km/s or 186,000 mi/s), which is the highest theoretical speed possible.

Science Fiction and Computer Game Mathematics

Space travel frequently appears in science fiction, where plots deal with various existing engineering or physics limitations. Hard science fiction is the more scientifically oriented subgenre, and it frequently includes extensions, discussions, and speculations dealing with the current scientific research. This tradition of blending science and literature started in the late nineteenth century with the works of Jules Verne; many of his then-fantastic devices and ideas (for example, televisions and submarines) were implemented relatively soon after.

As an example of experiments with scientific limits in literature, science-fiction spaceships may travel at superluminal (faster than light) speeds, often through non-physical spaces such as "hyperspace," "subspace," or "another dimension." These are terms from existing mathematical theories, which hard science fiction sometimes discusses.

Sci-fi spaceships may also be living organisms, completely or partially. This idea is a reflection of the current interest in bioengineering and has connections with exciting research in ecology, genetics, cybernetics, and artificial intelligence, as well as social sciences such as philosophy and bioethics.

Computer games and movies about space flight created a demand for applied mathematicians who can model fantastic situations with passable realism. The physics and mathematics of three-dimensional modeling is a fast-growing area, with new courses and programs opening in universities and an expanding job market. What started in the nineteenth century as an exotic occupation for very few writers has become a profession for many programmers and applied mathematicians.

Further Reading:
Battin, Richard. *An Introduction to the Mathematics and Methods of Astrodynamics.* New York: American Institute of Aeronautics and Astronautics, 1987.
National Aeronautics and Space Administration. "Design a Spaceship." http://www.nasa.gov/centers/langley/news/factsheets/Design-Spaceship.html.
Osserman, Robert. "Mathematics Awareness Month: Space Exploration." http://www.mathaware.org/mam/05/space.exploration.html.

Maria Droujkova

Spam Filters

Category: Communication and Computers.
Fields of Study: Number and Operations; Data Analysis and Probability; Problem Solving.
Summary: Spam filters use probability and Bayesian filtering to sort spam from legitimate e-mails.

Most people with an e-mail address receive unsolicited commercial e-mail, also known as spam, on a regular basis. Spam is an electronic version of junk mail, and has been around since the introduction of the Internet. The senders of spam (called spammers) are usually attempting to sell products or services. Sometimes, their intent is more sinister—they may be trying to defraud their message recipients. Since the cost of sending spam is negligible to spammers, it has been bombarding e-mail servers at a tremendous rate. Some estimate that as much as 40% to 50% of all e-mails are spam. The cost to the message recipients and businesses can be considerable in terms of decreased productivity and unwelcome exposure to inappropriate content and scams. As frustrating and potentially damaging as spam e-mail is, fortunately, much of it does not reach recipients thanks to spam filters. Spam filters are computer programs that screen e-mail messages as they are received. Any e-mail suspected to be spam will be redirected to a junk mail folder so that it does not clutter up a user's inbox. How does the filter decide which messages are suspect? Spam filters are implementations of statistical models that predict the probability that a message is spam given its characteristics. The filter classifies messages with large predicted probabilities of being spam, as spam.

Filters

Primitive filters simply classified a message as spam if it contained a word or phrase that frequently appeared in spam messages. However, spammers only need to adjust their messages slightly to outsmart the filter, and all legitimate messages containing these words would automatically be classified as spam. Modern spam filters are designed using a branch of statistics known as "classification." Bayesian filtering is a particularly effective probability modeling approach in the war on spam. Bayesian methods are named for eighteenth-century mathematician and minister Thomas Bayes. He formulated Bayes' theorem, which relates the conditional probability of two events, A and B, such that one can find both the probability of A given that one already knows B (for example, the probability that a specific word occurs in the text of an e-mail given that the e-mail is known to be spam); the reverse, the probability of B given that one knows A (for example, the probability that an e-mail is spam given that a specific word is known to appear in the text of the e-mail).

The underlying logic for this type of filter is that if a combination of message features occur more or less often in spam than in legitimate messages, then it would be reasonable to suspect a message with these features as being or not being spam. An extensive collection of e-mail messages is used to build a prediction model via data analysis. The data consist of a comprehensive collection of message characteristics, some of which may include the number of capital letters in the subject line, the number of special characters (for example, "$", "*", "!") in the message, the number of occurrences of the word "free," the length of the message, the presence of html in the body of the message, and the specific words in the subject line and body of the message. Each of these messages will also have the true spam classification recorded. These e-mail messages are split into a large training set and a test set. The filter will first be developed using the training set, and then its performance will be assessed using the test set. A list of characteristics is refined based on the messages in the training set so that each of the characteristics provides information about the chance the message is spam.

However, no spam filter is perfect. Even the best filter will likely misclassify spam from time to time. False positives are legitimate e-mails that are mistakenly classified as spam, and false negatives are spam that appear to be legitimate e-mails so they slip through the filter unnoticed. An effective spam filter will correctly classify spam and legitimate e-mail messages most of the time. In other words, the misclassification rates will be small. The spam filter developer will set tolerance levels on these rates based on the relative seriousness of missing legitimate messages and allowing spam in user inboxes.

Spam filters need to be customized for different organizations because some spam features may vary from organization to organization. For instance, the word "mortgage" in an e-mail subject line would be quite typical for e-mails circulating within a banking institution, but may be somewhat unusual for other businesses or personal e-mails. Filters should also be updated frequently. Spammers are becoming more sophisticated and are figuring out creative ways to design messages that will filter though unnoticed. Spam filters must constantly adapt to meet this challenge.

Further Reading

Madigan, D. "Statistics and the War on Spam." In *Statistics: A Guide to the Unknown*. 4th ed. Belmont, CA: Thompson Higher Education, 2006.

Zdziarski, J. *Ending Spam: Bayesian Content Filtering and the Art of Statistical Language Classification*. San Francisco, CA: No Starch Press, 2005.

BETHANY WHITE

Televisions

Category: Architecture and Engineering.
Fields of Study: Data Analysis and Probability; Geometry; Measurement; Number and Operations.
Summary: Innovations in television technology rely upon a sophisticated use of mathematics, physics, and engineering.

Humans process reality by initially recording light and sound waves through the eyes and ears and then transmitting these data to the brain where they are transformed and synthesized into intelligible matter. In a similar manner, the engineering challenge of television from its conception has been to record data, transmit them (via electricity), and then reconstitute them at a physical distance from its origin. Television is a relatively recent invention. The first appearance of the word (a combination of Greek and Latin words, meaning "far-seeing") occurred in 1900 at the International Electricity Congress at the Paris Exhibition. It was not one single person who invented television, but a number of scientists, engineers, and visionaries working independently in different countries who devised the necessary technology and mathematics. Television has changed dramatically from its first appearance as an electromechanical system to electronic systems, including cathode ray tube (CRT), liquid crystal display (LCD), plasma, and three-dimensional (3D) television.

Image Scanning and Aspect Ratio

Scanning the image required it to be disassembled into discrete pieces of picture that could then be transmitted separately and reassembled as a sequence of images on a screen, with each image recomposed from those smaller pieces of the picture. If the sequence of images could be displayed on the receiver's end rapidly enough, they would appear to the human eye as a continuous whole of moving images. This approach makes use of the fact that the human eye can distinguish two parallel lines only if they are about one-thirtieth of a degree apart and will blend 12 images per second into a moving whole. In the 1920s, the transmission of images went from an unacceptably choppy five per second to 12.5 and more.

The earliest scanning mechanism is known as the "Nipkow disk," named for the German physicist Paul

Nipkow, and versions and refinements of this were used as late as the 1930s. It consisted of a disk with a spiral of small holes in it and a photosensitive cell made of selenium on the other side of the plate from the image. One revolution of the disk corresponded to one complete image, with the holes as they rotated capturing the image in a series of lines. The number of such lines depended on the number of holes, which thus determined the degree of resolution of the image. A second disk was then rotated at the receiving end, playing back the captured image. One drawback of the Nipkow disk was that the scanned lines were not linear, which changed the geometry.

Historians debate why Thomas Edison chose to represent the geometry of television using the rectangular 4:3 aspect ratio, which indicates the ratio of the width to the height of the image. Some hypothesize that Edison chose this because the ratio approximates the golden mean while others assert that his motivation was to save money by cutting 70 mm film stock in half.

The Society of Motion Picture Engineers adopted this ratio in 1917 and it was standard for many years. The international standard for high-definition television was devised mathematically in 1980 by electrical engineer Kerns H. Powers. Powers analyzed the common aspect ratios in use at the time and normalized them to a constant area to fit them in a rectangle. When overlapped via their centers, they shared a common inner rectangle. He computed the geometric mean to obtain the 16:9 aspect ratio that continues to be the standard for televisions in the twenty-first century.

A uniform aspect ratio for television created another problem of how to capture the ratio on 35 mm film. Mathematical principles were used to develop lenses that were "anamorphic," which stemmed from the Greek words meaning "formed again." Ultra Panavision used counter-rotated prisms, Technirama used curved mirrors and reflection principles, and CinemaScope used a cylindrical lens. However, the lenses created distortion problems as compared to spherical lenses. In the twenty-first century, mathematics continues to play a role in anamorphic widescreen processes.

CRT Television

While electromechanical televisions such as the Nipkow disk were being developed, an electronic alternative that used a CRT rather than mechanical parts was also being explored. Philo Farnsworth and Vladimir Zworykin, among others, worked independently on this technology in the United States in the late 1920s. The diameter of the round picture tube, which was also the diagonal of the rectangular cover, was the critical parameter. Televisions are still measured on the diagonal in the twenty-first century.

The innovation involved harnessing electrical properties of matter. At the receiving end is a CRT—a glass vacuum tube, which receives the incoming transmitted signal that represents the picture, known as the "video signal" (audio and visual components are transmitted separately). At one end of the CRT is a cathode, which is heated so that it will radiate electrons (negatively charged particles) that are then attracted along the circuit to the other end of the tube (called the "anode end"), which is at positive electric potential for this purpose. This beam of electrons is focused electrically by charged plates and can be delicately manipulated by interactions with a magnetic field produced by electric current passing through coils.

3D Television

Stereoscopic effects produced by special televisions add a perceived depth of dimension to standard television that has previously been represented by only height and width, though this technology was still in its infancy at the start of the twenty-first century. Mathematics plays an important role in the evolution of home 3D technology. For example, it is used in determining the proper viewing distance and angle, which depend on the geometry of the display and the location of the viewer in a (often) small space. However, there is a great deal of variability in the process. At the start of the twenty-first century, some people complain of headaches caused by improper parallax, interocular distance in the images or display, or difficulty in interpreting the motion.

(Photos.com)

At this end of the tube is a photosensitive phosphor-coated screen, which has the property of responding to the beam of electrons by emitting light that is proportional in intensity, point for point, to the beam that is moved across it. The video signal is synchronized with the electron beam so that the variations in the beam relay image information. The beam moves line-by-line, lighting the phosphor that illuminates the screen on which the image is viewed. Color images necessitate a more complicated technology than black-and-white images: three signals, one for each of the primary colors (red, green, and blue) and three electron beams are exploited to produce color images.

LCD and Plasma

CRT television was standard through the 1980s but the line-by-line sweeping of the electron beam across the screen takes time and faster technology is available on high-definition television (HDTV), which depends on either an LCD or a plasma screen. The image received via these newer technologies is still comprised of small units, called "pixels" (an abbreviation of "picture elements"), but these operate differently. In an LCD system, each pixel is deployed by an electrically stimulated liquid crystal, which undergoes internal molecular rearrangement in such a way as to polarize (filter) light that is shone from the back. Intensity of light is adjusted by a blocking procedure similar to sunglasses. In a plasma screen, however, each pixel functions like a miniature fluorescent light, since it contains a mixture of gases and mercury that respond to electric charge by radiating energy that in turn causes phosphor on a screen to emit light.

Further Reading

Abramson, Albert. *The History of Television, 1880–1941*. Jefferson, NC: McFarland & Company, 1987.

———. *The History of Television, 1942–2000*. Jefferson, NC: McFarland & Company, 2003.

Noll, A. Michael. *Television Technology: Fundamentals and Future Prospects*. Norwood, MA: Artech House, 1988.

Todorovic, Aleksandar Louis. *Television Technology Demystified: A Non-Technical Guide*. Philadelphia: Elsevier, 2006.

Connie Wilmarth

Thermostat

Category: Architecture and Engineering.
Fields of Study: Algebra; Measurement.
Summary: Thermostats are mathematically calibrated according to physical principles to regulate temperature in a variety of settings.

Thermostats and thermometers are related instruments that perform different tasks. A thermometer measures ("meter") heat ("thermo") to determine and display a current temperature. On the other hand, a thermostat is designed to keep the heat ("thermo") stationary ("stat") to help maintain a desired temperature. Inventor and college professor Warren S. Johnson produced the first electronic room thermostats in 1883. He installed them in classrooms to keep students more comfortable in cold weather and to minimize outside interruptions. In the twenty-first century, thermostats are most commonly found inside vehicle engines and as a part of residential, commercial, or industrial heating systems—though they can also be found in appliances, like gas stoves.

Automobiles

In an automobile engine, the thermostat helps regulate temperature so that the engine operates properly and efficiently. The thermostat acts as a control valve for the coolant fluid, which flows within an engine and to a separate radiator that helps to cool the hot coolant. When an engine is first started, the thermostat is closed, and the coolant flowing within the engine cycles through only the engine until it warms up to an ideal temperature. The thermostat measures the temperature change using a special type of wax. Initially, the wax is solid but as the temperature of the surrounding coolant increases, the wax melts and expands to allow hot fluid to flow from the engine to the radiator and cooler fluid to flow from the radiator back in to the hot engine. If the engine gets too hot, the thermostat will open more to allow coolant from the radiator to permeate through the engine. On the other hand, if the engine begins to get too cold, the thermostat will begin to close, allowing less coolant into the radiator and more coolant to cycle through the engine to heat it back up. The thermostat is mathematically calibrated to the engine type and will automatically make the needed corrections as the vehicle is in use.

Buildings

A thermostat used to control temperature in a building similarly does not directly heat (or cool) the rooms. In this situation, it controls a heating (or cooling) unit, which is used to help regulate the temperature. In many systems, a bimetallic strip is used to measure the temperature of a room. Metals expand and contract as they heat and cool. Bimetallic strips work because different metals expand and contract at different rates. A strip of steel and a strip of copper (or brass) will be placed together and the ends secured to each other. If the temperature does not change, the strip remains flat. When the temperature changes, the different rate of expansion or contraction will cause the flat strip to develop a curve toward the metal that has changed less. The amount of curvature can be matched mathematically to a specific degree or range of change in temperature, triggering the system to adjust accordingly.

To increase the sensitivity of the thermostat, most bimetallic strips are long and coiled inside the thermostat. The coil loosens or winds more tightly with a change in room temperature. At a certain point, the bimetallic strip's movements will trigger the heating unit to turn either on or off. Once turned on, the thermostat uses weights or magnets to keep the heating unit from turning off too quickly. Without these devices, the thermostat would create short cycles (turning on and off quickly), which are generally inefficient and could cause a premature failure of the heating unit. Since the bimetallic strip's movement depends directly on the temperature of the immediately surrounding air, the thermostat should not be placed in a location that would cause an inaccurate reading. One common mistake is placing the thermostat by a heat register, where hot air flowing out will trigger the thermostat to turn the heating unit off before the rest of the room has acclimated.

Electronic Variations

More advanced thermostats frequently use electronic rather than electromechanical sensors and may have more than a simple on-off setting. Setpoint staging uses one type of heating process, or *stage*, when the room temperature is within two degrees of the thermostat setting and another when the difference is greater than two degrees from the thermostat setting. Time-based staging activates a secondary stage or unit after the first stage runs for a predetermined amount of time, indicating that the room is colder or hotter than some preset value. Multistage thermostats analyze variables such as the current room temperature, the desired temperature, and the amount of time it takes for a space to warm or cool one degree to determine mathematically when to use a second heating stage.

Other Thermostat Applications

The term "thermostat" is also used in statistical thermodynamics, which applies probability theory to systems made up of a large number of particles. This field of study helps relate the large-scale properties of materials observed by people in everyday life to the microscopic properties of the atoms and molecules from which they are made. Here, a thermostat mathematically maintains a constant temperature in computer simulations of molecular dynamics by realistically exchanging the energy of endothermic and exothermic processes that happen during the simulation. For example, the Gaussian thermostat, named for mathematician Carl Friedrich Gauss, maintains system temperature by rescaling the velocities of the simulated atoms at each individual step of the simulation.

Further Reading

Automatic and Programmable Thermostats. Merrifield, VA: Energy Efficiency and Renewable Energy Clearinghouse, 1997.

Brumbaugh, James E. *Audel HVAC Fundamentals, Heating Systems, Furnaces and Boilers*. 4th ed. Hoboken, NJ: Wiley, 2007.

Cleveland, Cutler J., et al. *Dictionary of Energy*. Expanded ed. Oxford, England: Elsevier, 2009.

Miles, Victor Chesney. *Thermostatic Control; Principles and Practice*. London: Newnes, 1965.

CHAD T. LOWER

Time Signatures

Category: Arts, Music, and Entertainment.
Fields of Study: Measurement; Number and Operations.
Summary: Musical time signatures are mathematically defined and are cyclical in nature.

A time signature is a musical notation that defines the meter of a particular composition or a portion of a composition. It establishes a hierarchical, cyclic relationship among beats and among the subdivisions of those beats, which are inherently mathematical in nature. The history of time signatures is somewhat unclear. Some suggest that time signatures first made their appearance around 1000 C.E., though they may not have looked like the ones used in the twenty-first century. Others date the development of the fractional-form time signature closer to the fifteenth century. Nearly all modern Western music uses time signatures or some type of grouped pulses. Along with tempo (rate of beats), musicians use time signatures to gain an understanding of the relation of the elements of a piece of music to one another in time, particularly with regard to a contextual temporal metric.

A time signature normally consists of two integers,

$$\frac{n}{b}$$

written with one directly above the other. Although it is often notated in prose as a fraction (for example, n/b), it is not a fraction and does not contain a dividing bar or solidus. A time signature appears in the first measure of a composition (in the staff following the clef and key signature), where it defines the default meter for the composition as a whole or until any subsequent time signature occurs that establishes a new default.

Meters and Beat

Time signatures may define various types of meters: simple, compound, complex, additive, or open. In simple meters (those in which the beats have a binary division), the upper integer indicates the number of beats in any one measure. The lower integer is conventionally expressed as a power of two, $b = 2^m$, and specifies what rhythmic value receives the beat. For instance, the time signature

$$\frac{2}{4}$$

indicates a simple meter in which every measure contains two beats and the quarter-note value is the relative duration of each beat. In compound meters (where beats divide into triples), the upper integer n, which is larger than three and divisible by it, designates that each measure contains $n/3$ beats. The lower integer, $b = 2^m$, indicates that the dotted $1/2^{m-1}$-th note receives the beat (the total relative duration of a $1/2^{m-1}$-th note and a $1/2^m$-th note). For example,

$$\frac{6}{8}$$

is the time signature for the compound meter in which each measure has two beats, and the dotted quarter-note duration (a quarter-note value plus an eighth-note value, or equivalently three eighth-notes) represents the beat.

Meters: Complex and Open

Complex meters incorporate beats that normally divide into a mixture of twos and threes. For example, the time signature

$$\frac{5}{8}$$

(each measure has the duration of five eighth-notes) might divide into two unequal beats: one with two subdivisions and one with three. The time signature for a complex meter might also be notated as an additive meter, wherein the upper value is actually an arithmetic expression that agrees with this pattern. For instance, the complex meter

$$\frac{5}{8}$$

could be indicated by the time signature

$$\frac{2+3}{8}$$

An open meter is notated by the symbol 0 in place of a more traditional time signature. It indicates that the duration of each measure is defined merely by the rhythmic values or graphic spacing of the notes it contains and does not incorporate a recurring or otherwise specified pattern of beats.

Cyclic Groups

Because of its cyclic nature, meter suggests a modular temporal space, similar to clock time. Algebraically, one might use cyclic groups to model different types of meters. The time signature is useful in determining the order of such a cyclic group, n from above, and what relative duration represents a generating unit, b from above. Then, the first beat of a measure, beginning at time-point zero, would associate with the identity element of the cyclic group, and so on through the nth beat of the measure. Any subsequent measures would represent additional cycles through these sequential group elements.

Interesting Time Signatures

Some time signatures are frequently used, like the lilting rhythm of the following:

$$\text{the waltz } \frac{3}{4} \text{ or the quick Sousa march } \frac{6}{8}.$$

A mathematician might argue that the number of time signatures is limited because the number of beats per measure quickly becomes divisible by a smaller number, making it a multiple of another time signature. However, in music theory, time signatures have a broader meaning in terms of tempo and musical phrasing, not just counts of beats. Interesting compositions have been constructed by considering the mathematical properties of time signatures. Robert Schneider of indie rock band The Apples in Stereo composed a score for a play written by mathematician Andrew Granville and his sister Jennifer Granville in which all the time signatures had only prime numbers of beats per measure. It also included Greek mathematics related to primes in musical form. An entire subgenre of music called *math rock*, which emerged in the 1980s, is typified by uncommon time signatures such as

$$\frac{13}{8} \text{ or } \frac{7}{8}.$$

These complex rhythms can also be found in some mainstream music, such as the song "Anthem" by Rush, which is partially written in

$$\frac{7}{8} \text{ time.}$$

Further Reading

Lewin, David. *Generalized Musical Intervals and Transformations*. New Haven, CT: Yale University Press, 1987.

Mazzola, Guerino. *The Topos of Music: Geometric Logic of Concepts, Theory, and Performance*. Basel, Switzerland: Birkhäuser, 2002.

Rastall, Richard. "Time Signatures." *Grove Music Online*. Edited by L. Macy. http://www.grovemusic.com.

Wright, David. *Mathematics and Music*. Vol. 28 of *Mathematical World*. Providence, RI: American Mathematical Society, 2009.

Robert W. Peck

Toilets

Category: Architecture and Engineering.
Fields of Study: Algebra; Geometry; Measurement.
Summary: The geometry of modern toilets has been analyzed by engineers using a variety of mathematical and statistical methods.

In all human societies, the disposal of bodily waste has been a primary health concern. It has been estimated that the average human being produces one to two liters of urine and one-quarter to one-half kilogram of feces each day. Fecal matter, in particular, can contribute to the spread of a wide range of diseases, as bacteria and other pathogens can enter food and water when waste is not treated properly. Such problems are especially prevalent in areas of high population density and limited water resources. Over time, a range of toilets and treatment systems have been developed to deal with sewage. Because of the lack of resources and infrastructure, many places in the world in the twenty-first century still contend with waterborne diseases that originate in human waste.

History

Given that many mammals, including most primates, choose to defecate in selected areas in their habitat, it is likely that humans have had specific defecation sites throughout history. Dry toilets, such as pit latrines and outhouses, are ways communities formalized the locations in which humans defecate and are still used in many parts of the world in the twenty-first century. In these systems, waste is concentrated in one place, ideally where it will not infect drinking water. The earliest sitting toilets that used running water to carry waste away date to at least 2500 B.C.E. in the civilizations of the Indus Valley, in what is now India and Pakistan. In 1596, Queen Elizabeth I's godson, Sir John Harrington, invented the first indoor flushing toilet. In 1775, Alexander Cummings, a Scottish watchmaker who studied mathematics, filed a patent for a flush toilet. However, it was not until the late 1700s in Europe and 1800s in America that further modifications and inventions ushered in an age of modern plumbing.

Design and Operation

The geometry of modern toilets is essential to their efficiency and is extensively analyzed by design engineers using a variety of mathematical and statistical methods. The modern home tank toilet consists of a storage tank, a bowl, and an *s*-shaped siphon. Water is stored in the tank. When the toilet is flushed, this water is released into the bowl through rim jets on the underside of the toilet's rim and through a tube called the "siphon jet" that allows most of the water to flow directly into the bowl. The bowl is attached to an "s"-shaped tube, and the influx of water from the tank into the bowl pushes the waste and water over the lip of the "s" and down to an attached waste system. The bowl clears because of the siphon-action created. When the toilet finishes flushing, air enters the siphon tube and stops the siphon. Meanwhile, a flapper valve in the toilet tank closes the connection between the tank and the bowl and allows the tank to refill.

New Developments

The flush toilet takes a large volume of water to operate. In an era of increasingly limited resources, there has been a movement to create low-flush and no-flush toilets. For example, toilets manufactured in the United States prior to 1994 used 13 liters of water per flush. The Energy Policy Act of 1992 required that toilets use six liters or less per flush, and as of 2011, high-efficiency toilets used 4.8 liters per flush. In Europe, dual flush toilets are common, providing the user with a choice of how much water to use depending on whether urine or feces is being flushed. Other technologies, including composting toilets that require no water and allow waste to biodegrade for use as fertilizer, have been developed for use by ecologically conscious consumers and people in areas of the world where water or sewage treatment facilities are limited. In addition, a number of toilets have been developed that include warmed seats, water and air jets for cleaning and drying the user, and built-in stool and urine analysis for health assessments.

Modeling Toilet Use

Many modern homes now have multiple toilets and ensuring adequate toilet facilities in public places requires planning and calculation. Two statistical studies of public-restroom use in the late 1980s are still referenced into the twenty-first century. They focused on the amount of time men and women spent in the restroom and they provided some of the first quantitative evidence that women take longer and thus require more toilets. This equity principle is known as "potty parity" and has been enacted into law in many places.

Further Reading

George, Rose. *The Big Necessity: The Unmentionable Subject of Human Waste and Why It Matters*. New York: Henry Holt and Co., 2008.

Raum, Elizabeth. *The Story Behind Toilets*. Chicago: Heinemann Library, 2009.

JEFF GOODMAN

Coriolis Effect

There is a frequently recurring question of whether the swirl of the water in toilets in the southern hemisphere is opposite that in the northern hemisphere. This notion has been perpetuated in many ways, including popular television shows and scientific programming or textbooks. It is true that large oceanic and atmospheric phenomena, such as hurricanes, will spin in opposite directions in the two hemispheres because of the Coriolis effect. In a small-scale system like a toilet, the geometry of the apparatus, along with water turbulence or temperature, is a much more important factor—a fact that has been verified through systematic experimentation.

Traffic

Category: Travel and Transportation.
Fields of Study: Algebra; Geometry; Problem Solving.
Summary: Mathematical models and statistical analysis of traffic flow suggest solutions.

Traffic flow is studied using mathematical and statistical techniques and computer simulations in order to better understand the movement of vehicles on roads and highways. Americans drive their vehicles almost 3 trillion miles per year on approximately 4 million miles of public roads. Mathematical models have shown that the behavior of even a single driver can have a broad impact on overall traffic flow in this dynamic system. As every driver knows, traffic patterns can often be unpredictable and frustrating, leading to driver stress, accidents, pollution, wasted fuel, and wasted time. Mathematical analysis of traffic congestion can provide transportation engineers with insights leading to improvements in efficiency and safety in the transportation of goods and people. A mathematical understanding of traffic flow patterns can also provide guidance for the design of roadways and provide more accurate calculations of trip itineraries and real-time driving times. These can be disseminated to the public and used in intelligent transportation systems.

The use of mathematics to describe traffic flow patterns slowly originated in the 1930s in order to study road capacity and also to begin to address traffic-related questions, such as how does traffic move through intersections. The mathematical investigations of vehicular traffic increased rapidly in the 1950s, mainly because of the expansion of the highway system after World War II. In the twenty-first century, theoretical models of traffic are utilized by high-performance computers, which can simulate the motions of vehicles on virtual road networks of entire cities and regions.

Traffic engineers distinguish between uninterrupted traffic flow situations (for example, traffic streams on highways and other limited-access roads) and interrupted flow circumstances (for example, where two or more traffic streams meet at a road intersection). The methods suited to analyze a particular traffic scenario depend on whether the flow is interrupted or uninterrupted. When formulating a mathematical description or model of traffic, one must attempt to account for the interplay between the vehicles and the drivers, the layout of the road system, traffic lights, road signs, and other factors.

Queuing theory, which is essentially the mathematical theory of waiting lines, is a probabilistic framework used for analyzing various traffic flow problems, such as optimizing vehicle passage through an intersection or traffic circle, calculating vehicle waiting times at tollbooths, and other similar waiting problems. On the other hand, car-following models and hydrodynamic modeling are deterministic approaches for analyzing traffic flow on long stretches of road.

Car-Following Traffic Models

Car-following models, also known as *microscopic models*, are considered from the point of view of tracking the movements of a line of $n = 1, \ldots, N$ individual cars driving in the same direction down a road in order to try to predict their exact positions $x_n(t)$, velocities $v_n(t)$, and accelerations $a_n(t)$. The starting point for car-following problems is to model how the driver of a car reacts when the vehicle directly in front of it changes speed (it is assumed for simplicity that there no passing is allowed). As a first crude estimation, one could assume a driver adjusts instantaneously according to the relative speed of the driver's car and the vehicle in front:

$$a_n(t) = C\left[v_{n-1}(t) - v_n(t)\right]$$

where C is a constant of proportionality, called the *sensitivity parameter*, which can be measured experimentally. A more realistic assumption would be that a driver adjusts with a lag response time of about one or two seconds, to a maneuver by the vehicle in front of it:

$$a_n(t) = C\left[v_{n-1}(t-T) - v_n(t-T)\right]$$

where T is the time lapse because of the driver's delayed reaction. Equations with delays such as these are then solved to keep track of each vehicle as the traffic moves. Numerous additional assumptions and effects have been incorporated into more sophisticated theories of car-following, such as considering the impact of spacing between cars, the effect of aggressive or cautious driving, and the effect of drivers looking ahead in the road and reacting to the motions of multiple vehicles in front of it.

Hydrodynamic Traffic Models

Hydrodynamic modeling, also called "continuum modeling," considers the flow of a traffic stream to be analogous to the flow of a compressible fluid in a pipe. Continuum traffic models do not keep track of the positions of individual vehicles, like car-following models, but track averaged, macroscopic quantities. For a long stretch of crowded road, such as an interstate highway, three important quantities of interest are flow rate (Q in vehicles per hour), vehicle speed (V in miles per hour), and vehicle density (ρ in number of vehicles per mile). These variables, of course, can vary along the stretch of road in both space and time, and their relationship is described algebraically as $Q = \rho V$. Furthermore, based on observations of traffic patterns over the years, it has been posited that for a given stretch of road, there exists a direct relationship between the flow rate and density. What has essentially been observed is that, on a road having some maximum flow rate, there is a critical vehicle density below which speed is not severely impacted but above which speed reduces. As the density continues to increase, then eventually flow rate reduces, and traffic becomes completely congested. For a concrete example, Greenshield's model postulates a simple linear relation between vehicle speed and density,

$$V = V_{free}\left(1 - \frac{\rho}{\rho_{jam}}\right)$$

where the parameter V_{free} is the free flow speed of a vehicle that is unencumbered, and ρ_{jam} is the density corresponding to bumper-to-bumper traffic. Then, the flow-density relation would be given by

$$Q = V_{free}\,\rho\left(1 - \frac{\rho}{\rho_{jam}}\right).$$

This parabolic function begins to capture some of the flow-density behavior that is observed on some real roads, although it is certainly an oversimplification. If the traffic density is zero ($\rho = 0$), then the flow rate must also be zero ($Q = 0$). Additionally, in bumper-to-bumper traffic ($\rho = \rho_{jam}$), the flow rate is zero, or very nearly zero in reality.

In the Lighthill–Whitham–Richards (LWR) theory of traffic, a long stretch of road is considered that has no entries or exits. On such a stretch of road, the number of vehicles must be conserved, and this fact combined with a flow-density relation gives rise to an equation, called a "conservation law," that predicts how vehicle density varies along the stretch of road. When a traffic jam occurs, it manifests as a sudden disturbance, or shock-wave, in the vehicle density along the road. LWR theory and other much more sophisticated continuum models of traffic can predict conditions under which traffic jams will form, propagate, and dissipate. Common reasons for traffic jams are accidents, construction, lane merges, and other changes in road capacity. However (as all drivers have experienced) sometimes "phantom jams" occur on highways for no apparent reason. These phantom jams can also be explained by continuum traffic models.

Further Reading

Daganzo, Carlos F. *Fundamentals of Transportation and Traffic Operations*. Oxford, England: Pergamon-Elsevier, 1997.

Gazis, Denos C. *Traffic Theory*. Norwell, MA: Kluwer Academic Publishers, 2002.

May, Adolf D. *Traffic Flow Fundamentals*. Upper Saddle River, NJ: Prentice Hall, 1990.

Anthony Harkin

Trains

Category: Travel and Transportation.
Fields of Study: Algebra; Geometry; Measurement; Number and Operations.
Summary: Trains and railways present interesting mathematical problems related to force and load, scheduling, and geometry.

Railroads influenced nearly every aspect of nineteenth and early twentieth century U.S. society. Companies building infrastructure for railroads (and railroads themselves) dominated the U.S. economy as more goods and people were transported via rail. Investors clamored to profit from the railway boom, inspiring engineers and mathematicians to improve the technology used in the railway system. As more people traveled by train, punctuality and reliability needed to improve. Time zones in the United States were established primarily because competing rail companies used different standard times for their schedules. In addition, Christophorus Buys-Ballot and others conducted experiments using trains to explore the Doppler effect, named for mathematician and physicist Christian Doppler. At the start of the twenty-first century, wooden and electric railway sets remain popular toys with children of all ages, while railroad enthusiasts design elaborate model train layouts in various scales reflecting the days when towns were centered around train stations.

Locomotives

Locomotives are classified using the Whyte system, named for mechanical engineer Frederick Whyte, which utilizes numbers to describe the wheel arrangement of the engine. For example, a 4-8-4 type locomotive has four wheels in the front, 8 driving wheels in the middle, and 4 wheels in the rear. The capacity of a locomotive depends on the amount of friction the driving wheels have with the track and the weight of the engine over the driving wheels. These quantities are related by the equation $F = MW$, where F represents the maximum pulling force of the train, M represents the coefficient of friction between the wheels and the track, and W is the portion of the weight of the locomotive over the driving wheels. While this relationship indicates that heavier trains can pull larger loads, more power is needed to move the train, leading to higher fuel costs. Increasing the coefficient of friction gives the train better traction and thus more pulling force, so most locomotives have a sandbox on the front from which sand is sprayed onto the track when the rails are slippery. Though friction is needed to get the train started, reducing M increases efficiency once the train is in motion, lowering operating costs.

Modern diesel-electric locomotives use high-tech designs to achieve more horsepower while reducing engine weight significantly. Equipped with a sophisticated array of sensors, onboard computers, and control systems, twenty-first-century trains maintain their hauling capacity while reducing fuel consumption and emissions. The future may see more magnetic levitation (Maglev) trains, which use magnetic fields to suspend the train above the track. The first commercial Maglev train opened in 1984 in Birmingham, United Kingdom, but ceased operations in 1995 in part because of design problems. A Maglev train in Japan recorded a maximum speed of 581 kilometers per hour (361 miles per hour) in 2003, the highest ever speed for a Maglev transport.

Passengers and Timetables

Commercial trains, whether passenger trains or freight trains, follow carefully written schedules. Composing these intricate timetables is a daunting task. Railways must ensure that trains do not collide on the tracks, and that goods and people are transported in a timely and efficient manner. In 2006, the Netherlands introduced a new railway timetable for all trains and mathematical modeling played a key role in developing the timetable. To determine how a set of trains should be routed through a station, researchers listed all feasible routes through the station for every train. Each combination of a train and a feasible route is represented by a node on a graph. Nodes on this large graph are connected if they belong to the same train or if there is a routing conflict between the train/route combinations. Presenting the scheduling problem in graph form enables sophisticated computer programs to generate a usable timetable. Additional modifications improve the efficiency of the timetable in the case of unexpected delays.

Railway passengers expect trains to be on time and to have sufficient space for a comfortable ride. Timetables can be fine-tuned to meet these customer demands using another type of mathematical modeling called "peak load management." Consultants work with railways to determine when trains are the most crowded and when passenger demand is highest. Mathematicians quantify the notion of "attractiveness," a measure of how satisfied a rider on a given train will be as a function of the journey time on the train, the time the passenger would like the train to arrive at its destination, and the actual arrival time. Another constant is added to the equation to determine how much attractiveness is reduced for each minute the actual arrival time differs from the customer's ideal arrival time. More terms can be added to measure the crowding on the train—overcrowding having a significant impact on attractiveness. Using this model, railways can develop timetables that increase the probability that a customer will ride on an "attractive" train. Further refinements to the model attempt to minimize the chance that a passenger will need to stand while riding.

Trains as Teaching Tools

Creative elementary school teachers have devised ways to use the appeal of toy trains to teach addition and subtraction. A colorful cardboard train is taped to a bulletin board and children count the number of cars on the train. Train cars are easily removed or added and the students see addition and subtraction in action by counting the number of cars on the new train. Wooden railway systems with magnetic couplings between cars also allow for easy joining and separating, making these toys excellent mathematical manipulatives when working with small groups of children. Older students may encounter the Two Trains puzzle. Two trains are on the same track traveling toward one other at a constant speed. A fly starts on the front of one train and flies toward the other train at a constant speed, faster than either train. Once the fly reaches the other train, the fly immediately turns around and continues buzzing back toward the first train. How far does the fly travel before being smashed when the two trains collide?

(Photos.com)

Track Geometry

Freight yards use combinations of switches, sidings, and turnaround loops to sort railway cars, assembling them into trains bound for various destinations. The fact that trains cannot pass each other on a single track leads to many challenges. The optimal arrangement of freight cars in the most efficient manner is another problem for mathematical modeling, but these fascinating switching systems have inspired mathematicians to investigate interesting questions involving train track layouts and railway switching puzzles.

A switch (also known as a "turnout," or "point") is a Y-shaped structure used to split tracks into two lines or to combine two lines into one. The directional nature of a switch makes the dynamics interesting: trains entering at the "top" of the Y will always exit through the bottom branch, but trains entering through the bottom have the option of traveling on the left branch or the right branch. Switches are used to sort cars in freight yards, enable locomotives to move onto a siding to allow a train traveling the opposite direction on the track to pass, and make it possible via a turnaround loop for a train traveling one direction to reverse direction.

How can two trains traveling in opposite directions, say eastbound and westbound, pass one another? If there is a siding long enough to contain one of the trains, the problem is easy. But what if only one car can occupy the siding at a time? Variations on this train-passing puzzle have been around for over a century. The trains can still pass each other through clever use of the siding. The eastbound train leaves its cars behind, moves onto the siding, and waits for the westbound train to pass through. After the eastbound engine emerges from the siding, the westbound train backs through the siding, bringing along one of the eastbound train's cars and leaving that car on the siding. After the westbound train has pulled forward past the siding, the eastbound train can pick up its car, and the process repeats until the entire eastbound train is through.

Imagine a child playing with a toy railroad. Given a set of switches and plenty of track, how many different layouts can the child make? To determine whether two track layouts are different, the structure is transformed into a graph, with nodes representing lengths of track. Nodes are connected if there is a switch allowing a train to travel from one length of track to another. Layouts are said to be different if their graphs are the same. A child with two switches can make five distinct layouts. Using more switches and combinations of other types of switches, like the three-way pitchfork-shaped variety, even more layouts can be made and counted using mathematics.

Further Reading

England, Angela. "Train Math Lesson Plan." http://www.suite101.com/content/train-math-lesson-plan-a45144.

Gent, Tim. "Model Trains." http://plus.maths.org/content/model-trains.

Hayes, Brian. "Trains of Thought." *American Scientist* 95, no. 2 (2007).
Kroon, Leo. "Mathematics for Railway Timetabling." *ERCIM News* 68 (2007).
Lynch, Roland H. "Locomotives." *Ohio State Engineer* 23, no. 5 (1940).
Peterson, Ivars. "Ivars Peterson's MathTrek: Laying Track." http://www.maa.org/mathland/mathtrek_01_08_07.html.

Mark R. Snavely

Traveling Salesman Problem

Category: Travel and Transportation.
Fields of Study: Geometry; Problem Solving.
Summary: The traveling salesman problem is a notable applied mathematics problem that is simply constructed and may be unsolvable.

Imagine a salesperson that needs to travel to 30 cities. The salesperson wants to begin in his or her hometown, visit every city exactly once, and return to the hometown. In what sequence should the salesperson visit the cities in order to minimize the total amount of traveling time on the road between cities? The significance of the traveling salesman problem (TSP) lies in the fact that many other problems can be translated into a traveling salesman formulation and that a brute force check-all-the-possibilities approach will take prohibitively long—even for moderately sized problems (like the example) and with the use of fast computers.

Many problems can be translated to the TSP. The travel time between cities can be replaced by distance, cost, or other measures. Hence, in essence, this problem captures many sequencing problems where a number of tasks have to be sequenced and the costs can be modeled appropriately. Problems as diverse as optimizing the routes of garbage trucks, planning the sequence of motions performed by a robot, and ordering genetic markers on a chromosome have been modeled by the TSP.

Solving the TSP

Why is solving the TSP hard? If one decides to solve the problem by checking all the possibilities and then choosing the best one, then the sheer number of possibilities will make the problem impossible to solve. For example, with 30 cities and starting at a hometown, initially there are 29 cities to choose as a first destination. Regardless of the first choice, there are 28 cities to choose from next and so on. The total number of possible ways to start from a hometown, traverse each of the 30 cities exactly once and return to the hometown is

$$29! = 29 \times 28 \times \ldots \times 3 \times 2 \times 1$$
$$= 8,841,761,993,739,701,954,543,616,000,000$$
$$\approx 8.8 \times 10^{30}$$

possibilities.

Even if a computer checked a million possibilities per second, checking all the possibilities would take more than 200,000,000,000,000,000 years—much longer than the age of the universe. Making the computer twice or 10 times faster still will not be enough to make the problem worth attempting.

Solution Through Algorithms

Could there be clever algorithms that solve the TSP faster? The TSP is among the problems that computer scientists call *NP-hard*. Given any algorithm for solving the TSP, certainly the number of steps needed by the algorithm grows as the size of the problem—namely the number of the cities—grows. If the number of steps in an algorithm as a function of the size of the problem is a polynomial, then it is generally believed that the problem is tractable. In other words, if there is one such polynomial time algorithm, then one can hope to find other more efficient ones and be able to solve even large-sized problems efficiently. At the start of the twenty-first century, it is not known whether the TSP has such a polynomial time algorithm. But it is known that if there is such an algorithm, then there is also efficient algorithms for a host of other problems of interest to computer scientists. For many years, researchers have looked for such algorithms and have not been able to find one, and the strong prevailing opinion is that no such algorithm exists (this is the famous $P \neq NP$ problem).

Even though the TSP is a difficult problem to solve in general, progress has been made in developing algorithms that do much better than the brute force method. In fact, very large instances—for example, one with 85,900 cities—of the TSP have been solved exactly. On another front, many approximation algorithms have been devised. These algorithms do not aim to find the absolute best solution but rather find a solution that is close to the best one. A simple approximation algorithm using minimum spanning trees, for example, can find a solution that is guaranteed to be no worse than twice the optimal solution. More sophisticated algorithms can find a solution within a few percentages of the optimal solution for a problem with the number of cities in the millions.

Further Reading

Applegate, David L., Robert E. Bixby, Vašek Chvátal, and William J. Cook. *The Traveling Salesman Problem: A Computational Study.* Princeton, NJ: Princeton University Press, 2006.

Gutin, Gregory, and Abraham P. Punnen. *The Traveling Salesman Problem and its Variations.* Dordrecht, The Netherlands: Kluwer Academic Publishers, 2002.

Shahriar Shahriari

Tunnels

Category: Architecture and Engineering.
Fields of Study: Algebra; Geometry; Measurement; Number and Operations.
Summary: Tunnels have long presented interesting mathematical and engineering problems.

A tunnel is a connecting passageway through materials like rock, earth, or water. Tunnel engineers must take into consideration issues like seepage and weight. Scientists and mathematicians create mathematical models of tunnels to investigate aspects like aquifers and safety issues. Analytic and closed form solutions are useful in engineering. Mathematical fields like graph theory, differential equations, geometry, probability, and trigonometry are important for modeling and measuring tunnels.

Mathematically Challenging Tunnels

Five centuries after it was completed, Hero of Alexandria gave a theoretical explanation that may explain how the Tunnel of Samos was constructed. Mathematical physicist Renfrey Potts had an undergraduate degree in mathematics. He worked as a consultant for General Motors and created car-following models. This work led to experiments on a testing track with just two cars that successfully predicted the optimum speeds for congested traffic in the Holland Tunnel in New York, named for engineer Clifford Holland. The Channel Tunnel between England and France represented a significant engineering and mathematical challenge. At the time of its building and into the twenty-first century, it had the longest undersea length of any tunnel in the world. It presented significant challenges including problems related to the topology and geology of the rock through which it was bored; significant water pressure; ventilation; communication; and the fact that construction was started at the same time from both ends, requiring exceptional precision to meet in the middle. This tunnel serves as a model for other underwater tunnel projects and many teachers use it to present mathematics concepts. Scientists and mathematicians also experiment with digital and physical wind tunnels as well as quantum tunnels.

Ancient Tunneling

The problem of delivering fresh water to large populations has been an ongoing human endeavor since ancient times. In the sixth century B.C.E., a one-kilometer tunnel was dug through a large hill of solid limestone to bring water from the mountains to the main city on the island of Samos. The Eupalinian aqueduct on Samos was designed by the ancient Greek engineer Eupalinos of Megara. The tunnelers worked from both ends and met in the middle, with an error less than 0.06% of the height. To achieve this remarkable result, Hero of Alexandria theorized that the tunnelers used a method based on similar triangles in order to determine the correct direction for tunneling. Mathematicians and scientists continue to debate the pros and cons of various theories of how this engineering marvel was constructed.

Modeling Tunnels

Tunnels can be modeled using coordinate geometry and equations. For example, knowing the height and width of a parabolic tunnel, one can determine the tunnel's height at different distances from the base center. To solve this problem, one needs to find the equation for the parabola choosing convenient *x-y* axes.

Frictionless Tunnels

The possibility of mathematical modeling allows for innovative and challenging ideas. What if a frictionless tunnel would be bored through Earth's center? Paul Cooper, a mathematician fond of Jules Verne's books, tried to answer this question in an issue of the *American Journal of Physics*. He set up and solved by computer a set of differential equations for tunnels that would provide minimum gravity-powered travel time between any two cities on Earth.

According to Cooper's differential equations, by freefalling in airless, frictionless, straight-line tunnels, passenger vehicles powered only by the pull of gravity could theoretically travel between any two points on the Earth's surface in a total time of only 42.2 minutes.

Accelerated by the force of gravity on the first half of the trip, the vehicle would gain just enough kinetic energy to coast up to the other side of the Earth. However, significant obstacles make such a project impossible in the twenty-first century. Subterranean temperatures reach extremes, even for relatively shallow tunnels of only a few miles deep, requiring huge cooling systems for vehicles. Also, it is almost certainly impossible to create a completely frictionless path without a rail or track of some type, leaving the vehicle with insufficient kinetic energy to complete its trip without a source of additional power. Consequently, such a tunnel is still science fiction more than science.

Further Reading

Apostle, Tom. "The Tunnel of Samos." *Engineering and Science* 1 (2004).

Cooper, P. W. "Through the Earth in Forty Minutes." *American Journal of Physics* 34, no. 1 (1966).

If the height and width of a parabolic tunnel are known, one can determine the tunnel's height at different distances from the base center by modeling using coordinate geometry and equations. (Photos.com)

Lunardi, Pietro. *Design and Construction of Tunnels: Analysis of Controlled Deformations in Rock and Soils.* New York: Springer, 2008.

Oxlade, Chris. *Tunnels.* Portsmouth, NH: Heinemann-Raintree, 2005.

FLORENCE MIHAELA SINGER

Universal Language

Category: Space, Time, and Distance.
Fields of Study: Communication; Connections Representations.
Summary: Mathematics has been proposed as a universal language; attempts have been made at a mathematics notation that would be recognizable on any planet.

From the beginnings of humanity, people needed to establish connections. Along with speaking, counting developed from the early stages of human evolution. Numbers and counting were necessary in the first civilizations to describe ownership, for trade, or for calculating taxes. Shapes and measures were needed to make furniture, buildings, and ritual places, as well as in landscaping, time-keeping, sky-charts, and calendars. Mathematics is present everywhere in the real world: in science, art, entertainment, business, and leisure. People use mathematics to describe the universe, and mathematics is commonly referred to as the "language of science or the universe." Albert Einstein questioned:

> At this point an enigma presents itself, which in all ages has agitated inquiring minds. How can it be that mathematics, being after all a product of human thought which is independent of experience, is so admirably appropriate to the objects of reality? Is human reason, then, without experience, merely by taking thought, able to fathom the properties of real things?

Some take this idea a step further and view mathematics as a universal or interstellar language or explore the creation of a universal language.

Debate

Those who consider that mathematics is a universal language reason that because mathematics arises naturally and humans possess the ability to be literate in the shared language of mathematics then it must be universal. Others criticize this viewpoint and note that learning mathematics is challenging for many people. Some scientists and mathematicians point to the fact that despite differences between cultures and natural languages, the discoveries in mathematics are the same all over the world because mathematics is so well-suited to describe reality. Discoveries that were simultaneous, like the formulations of calculus by physicist Sir Isaac Newton and mathematician and philosopher Gottfried Leibniz, appear to give even more credence to this viewpoint. However, Newton and Leibniz were able to share ideas and build upon the contributions of the same earlier mathematicians and they developed different mathematical approaches and terminology. In some examples of simultaneous discoveries, like for mathematicians in the Soviet Union and the United States, the researchers were quite separated. Other philosophers and mathematicians assert that humanity invents mathematics and distorts reality in accepting its postulates.

Physicist Werner Heisenberg's uncertainty principles seem to give rise to questions about whether anyone can objectively measure or quantify reality. Attempts to model the universe on a quantum and grand scale have led to both calls for and rejection of a theory of everything.

Creating a Universal Language

Scientists, mathematicians, philosophers, and linguists have long contemplated a language that is universal. Linguists explore languages for commonalities, and Search for Extraterrestrial Intelligence (SETI) researchers analyze signals for mathematical patterns. Some visual or graphical representations are also viewed as universal. In *De Arte Combinatoria*, Leibniz imagined

> . . . a general method in which all truths of the reason would be reduced to a kind of calculation. At the same time this would be a sort of universal language or script . . . for the symbols and even the words in it would direct the reason . . . It would

be very difficult to form or invent this language or characteristic, but very easy to understand it without any dictionaries.

Leibniz cited earlier attempts at universal languages, such as correspondances that converted words into numbers by physician Johann Becher or scholar Athanasius Kircher. George Dalgarno had invented a system for translating numbers into words. In 1678, Leibniz also developed this type of system: 81,374 would be written and pronounced as *mubodilefa*. For Leibniz, the digits 0–9 became the first nine consonants of the alphabet and powers of 10 were represented using vowels. Leibniz also planned to explore the logical foundations of geometry via a universal language but he did not continue this work.

Philosopher Sundar Sarukkai noted that: "The search for 'universal' language or 'pure' language is part of human history in all civilizations. In part, this reflects an enormous distrust of ambiguity in meaning." However, he also asserts that, "it is semantic ambiguity that allows individuals and societies to develop and flourish."

Further Reading

Ballesteros, Fernando. *E.T. Talk: How Will We Communicate With Intelligent Life on Other Worlds?* New York: Springer, 2010.

Jeru. "Does a Mathematical/Scientific World-View Lead to a Clearer or More Distorted View of Reality?" *Humanistic Mathematics Network Journal* 26 (June 2002).

Rutherford, Donald. "The Logic of Leibniz by Louis Couturat, Chapter 3 Translation." http://philosophyfaculty.ucsd.edu/faculty/rutherford/Leibniz/ch3.htm.

Sarukkai, Sundar. "Universality, Emotion and Communication in Mathematics." *Leonardo Electronic Almanac* 11, no. 4 (2003).

Yench, John. *A Universal Language for Mankind*. New York: Writers Club Press, 2003.

SIMONE GYORFI

Vending Machines

Category: Architecture and Engineering.
Fields of Study: Algebra; Geometry.
Summary: Ubiquitous vending machines use algebra and Boolean logic to function.

Vending machines are finite state machines, also known as "automata," that transition between states based on customer input data, such as product selection. Vending machine designers use mathematical models and Boolean algebra to determine the states the machine should transition into based on input data variables, with the outcome often expressed as a table. The control unit reads the data as either "true," meaning the machine recognizes the input language, or "false," meaning that it does not.

The first documented vending machine, invented by the Egyptian mathematician Hero of Alexandria, appeared c. 215 B.C.E. By the twentieth century, vending had developed into a billion dollar industry, and vending machines dispensed a variety of products. Older vending machines relied on the mechanical activity of knobs or levers activated by the customer to dispense the desired product. Vending machine operators utilize mathematics to determine potential and actual expenses and profits, as well as to process sales and stock data. For example, net income can be determined through the simple formula: Net Income = Income − Expenses.

Modern vending machines, however, utilize basic computing system processors to analyze customer input data, such as a letter and number, that corresponds to the desired product, which is then electronically dispensed. Modern advances in vending machine technology include card validators for debit and credit cards; voice activation; electronc message displays for insufficient funds, lack of change, or sold out products; and remote wireless diagnostics and data collecting to alert venders of the need for restocking or repair.

Vending machine control units are part of a class of abstract machines known as "finite state machines" or "automata"; in particular, they are deterministic or discrete finite state automata (DFA). Finite state machines are always in a position known as a "state," transitioning between these states based on input data. Designers use mathematical models in the design of finite state machines, such as vending machines. The machines are

A woman buying a beverage on a Tokyo street. Vending machines are extremely popular in Japan and there are machines that sell ramen noodles, alchoholic beverages, fruit and vegetables, batteries, and even clothing. (iStockphoto)

designed to recognize a regular language, converting computation into language recognition. Each state is labeled either "true" (accept the data) or "false" (reject the data) based on whether the machine recognizes the language of the input data.

Vending machine design utilizes Boolean logic or algebra, or algebra based on two logical values, in this case the values of "true" and "false." The general Boolean function is expressed through the formula

$$y = \sum (x, \ldots)$$

where $(x, \ldots)$ is equal to a set of Boolean variables with the values "true" or "false." Diagrams of the various states of the vending machine and the possible transitions between them can be converted into Boolean operations.

The control unit reads each string of input data, generally input from the vending machine customer, such as the diameter, thickness, or number of ridges of coins followed by product selection codes. Transition functions tell the machine which state it should enter based on input data. Transition functions are often represented in tabular form. The control unit changes its state with each data string entered until the final input, after which it outputs either "true" or "false" based on its final state. Vending machines also use the algebraic relationship between range and domain, where the range is the machine's output and domain is the customer's input. For example, a customer must input an equal or greater amount of money than the cost of the desired product.

Further Reading

Hopcroft, John E., and Jeffrey D. Ullman. *Introduction to Automata Theory, Languages, and Computation*. Reading, MA: Addison-Wesley, 1979.

Salomaa, Arto. *Computation and Automata*. New York: Cambridge University Press, 1985.

Salyers, Christopher D. *Vending Machines: Coined Consumerism*. Brooklyn, NY: Mark Batty Publisher, 2010.

Segrave, Kerry. *Vending Machines: An American Social History*. Jefferson, NC: McFarland & Company, 2002.

Marcella Bush Trevino

Video Games

Category: Games, Sport, and Recreation.
Fields of Study: Algebra; Data Analysis and Probability; Geometry.
Summary: Video games use the mathematical concepts of algorithms, matrices, and random numbers as part of their programming.

Video games are pervasive in modern society, from computers to television-based systems to applications that can be downloaded easily onto cell phones. There is an ongoing debate over what should be called the first video game. The narrower definition is a game generated by a computer and displayed on a video device. Others consider it to be any electronically based game displayed with video output. The most likely candidate is a 1940s invention by physicists Thomas Goldsmith and Estle Ray Mann. Their "Cathode-Ray Tube Amusement Device" was inspired by World War II radar displays and allowed the player to shoot virtual missiles at targets. Though patented, at the time it was too costly to produce commercially and only a few prototypes were ever made. Much of the mathematics used to design and operate computers also applies to video games and the various fields and professions are closely connected. Video game design programs offered by many colleges emphasize physics and mathematics education along with computer programming, as these skills are necessary to represent the real world in increasingly realistic ways. The new generation of body-sensing game controllers uses optics to detect a player's motion in three axes and translate it to corresponding movements within the game environment. While most people think of video games as entertainment, they are increasingly being incorporated into the classroom and other learning applications. In 2009, U.S. President Barack Obama initiated a campaign called Educate to Innovate, which seeks to use interactive games, among its other strategies, to improve the mathematical and scientific abilities of American students.

Simple Modeling Using Polygons

Any video game that has graphics needs to have a way of drawing a picture on the screen. A very basic program can take a turtle (or curser) on a screen and move it forward and rotate its direction clockwise. Many geometrical shapes are easy to draw using a turtle. For example, to tell the turtle to draw a rectangle, a simple program might tell the turtle to move 100 steps (which could be measured by pixels on the screen), turn 90 degrees, move 50 steps, turn 90 degrees, move 100 steps, turn 90 degrees, and move 50 steps. At this point, a 100 × 50 rectangle has been drawn, and the turtle is perpendicular to the position where it started.

A circle (or any object with a curve) would be much more difficult to draw using these commands because of the thickness of a pixel and the fact that the turtle cannot move half degrees. A user could try to tell the turtle to move one step then turn one degree. After repeating those commands 360 times, the turtle will be back where it began, and will have drawn a circle that is slightly less than 115 steps across. Technically, it did not draw a circle, but rather a polygon with 360 sides. A slight modification may be to tell the turtle to move two steps then turn one degree. After 360 repetitions, the turtle will appear where it started and the shape appears to be a circle that is twice as wide as the first shape drawn, about 229 steps wide.

There is a big gap between 115 steps and 229 steps wide. If a programmer needs a circle between those dimensions (or beyond those dimensions), the programmer can use mathematics to adjust the step length to get a circle of the desired size. The length across a circle is called the "diameter" and the distance around a circle is called the "circumference." The relationship between these two measurements is $C = \pi d$, where C is the circumference, and d is the diameter.

Since the turtle will be tracing the outside of the circle, it will travel the length of the circumference. The turtle will also be making 360 turns during its travel. Since each step should be the same length, one can find the length of each step by taking the circumference and dividing by 360. Since π is approximately 3.14, one can estimate the length of the step by multiplying 3.14 and the desired diameter and then dividing by 360.

Depending on the video game being created, a programmer will probably desire to draw more than circles and polygons. Using the above steps for a circle but only repeating the steps 180 times will yield a half circle, which could approximate the shape of a setting sun, the top of a silo, or the ice cream in a cone. More complex shapes, like drawing a long-haired cat, could be made by the turtle but the programmer now has a time concern. The programmer creating the directions to draw the cat and the fur on the cat would require a long time to type in the programming for the cat—and even more if the cat is supposed to move—since the repeat step would be used sparingly, if at all. On the users end, a large program with a lot of steps would take a long time to draw, depending on the speed of the computer or gaming system on which it is to be played.

Although video games are displayed on a two-dimensional screen, programmers now commonly create elements of the game in three dimensions. To mimic the body of an object, programmers create the outer shell of the object using a mesh of triangles or quadrilaterals. Depending on the detail desired, more meshes could be created. Once the object is created, it needs to be displayed on the screen. This process involves using a "point-of-view camera," which will change how the object is drawn based on where the camera is and how far away it is from the object. The triangle mesh of the object is adjusted accordingly. For example, as the object approaches the camera (gets closer to the screen), the triangles will elongate and become larger. A programmer that wants the object to get closer to the camera and rotate will use vectors and matrices (linear algebra) to adjust the size and the dimensions of each triangle in the meshes. Once the computer does the mathematical calculations to modify the triangle mesh, the point-of-view camera creates a two-dimensional image of the three-dimensional mesh in the orientation it has been set to. This two-dimensional image then gets projected to the viewing screen.

A game programmer working with multiple monitors. The screen on the right shows a portion of the large amount of source code used. (iStockphoto)

Interesting geometry is also found in the movement of objects through the game. In some cases, like the games Portal and the older PacMan, players can exit the playing field on one side of the screen and return from another side or in a different orientation. This property involves concepts like a torus and higher-dimensional analogs.

Color

When programming colors (assuming the screen is not monochrome), a programmer needs to remember that the primary colors for light are different than the primary colors for pigment. When drawing on paper, the three colors magenta (red), cyan (blue), and yellow can be combined in such a way as to create almost any other color. For example, many color printers only use three colors to print. Since most screens work based on a projection of light (whether a computer monitor or a television screen), the primary colors of light must be used. For light, the colors red, blue, and green are the primary colors; with these colors, any other color can be created. All three together make white, and no light at all makes black.

When coding colors, each of the three primary light colors is given an intensity value 0 – 255. This value is then converted to a two-digit hexadecimal number, where 00 is the decimal number zero and FF is the hexadecimal number 255. The hexadecimal number 12 would be an intensity level of 18. The hexadecimal number A0 would be an intensity level of 160. The programmer then takes these three intensity numbers and combines them to make a six-digit color number by placing the intensities in order for red, then green, and finally blue. For example, pure red would be FF0000 (intensity 255 for red and intensities 0 for both green and blue). Similarly, 00FF00 would be pure yellow and 0000FF would be pure blue. The color white would be represented FFFFFF (a combination of all three colors), whereas black would be 000000 (no light whatsoever).

Random Number Algorithm

Many video games that have been created offer a storyline or, at least, a progression to get from one stage or level to the next. Moving to the next level often requires a certain level of skill or collecting certain objects. On the other hand, there are video games that are created, like video poker or Tetris, where skill alone is not enough to do well. There is a certain random element that will determine the outcome. However, computers are not capable of creating random numbers. Instead, the video game console is pre-programmed with a list of pseudo random numbers. For example, every TI-84 calculator that has its memory reset will create the number 0.94359740249213 as its first "random" number. Obviously, if everyone obtains the same result, it cannot be random.

Using the TI example, every random number it produces will be a decimal between zero and one. If the game requires a number higher than one, the programmer merely multiplies the random number times the highest number they desire. For example, in Tetris, there are seven tetrominoes that could be selected for the next drop. A programmer may want a random number generated to determine the shape of the next piece. As a result, the programmer would create a random number, and then multiply it by seven to get a number between zero and seven; however, this number is still a decimal. The computer programmer can then tell the console to truncate the number, which would ignore everything beyond the decimal point giving an integer between zero and six. A final (optional) step would be to add one to this truncated integer resulting in a number between one and seven. Each block would get assigned a number, and the pseudo-random number that resulted would select the next block.

Further Reading

Dunn, Fletcher, and Ian Parberry. *3D Math Primer for Graphics and Game Development*. Plano, TX: Worldware, 2002.

Egan, Jill. *How Video Game Designers Use Math*. New York: Chelsea Clubhouse, 2010.

Flynt, John P., and Boris Meltreager. *Beginning Math Concepts for Game Developers*. Boston: Thomas Course Technology, 2007.

Chad T. Lower

Water Distribution

Category: Architecture and Engineering.
Fields of Study: Algebra; Data Analysis and Probability; Geometry; Measurement; Number and Operations; Problem Solving.

Summary: Mathematicians have long studied issues related to optimizing water distribution.

Water distribution has two separate but interrelated meanings: the natural physical distribution of water in the world and the way in which people choose to distribute available water. In some regions, accessing and distributing fresh water for human needs, like drinking and irrigation, can be a significant challenge. Roughly 70% of the Earth's surface is covered with water but most is saline (salty). Much of Earth's fresh water is in glaciers or underground. Some is polluted from human activities. In the early twenty-first century, approximately 20% of Earth's population lived in areas with insufficient fresh water because of climate or geography. About the same number lived in areas in which water existed but where technological or economic barriers limited effective distribution. Many systems have been devised throughout history and in different societies to access and distribute water. It is so valuable a resource that armed conflicts been fought over water. Mathematicians, scientists, and others who work on water distribution problems use mathematical techniques to design, build, optimize, and monitor water distribution and associated wastewater systems. For example, graph theory is used to model water distribution networks. Graph edges may represent pipes and nodes represent intersections, junctions, and access points. Statistical and topological methods can be used to compare networks in terms of capacity and reliability against failure.

Irrigation

Irrigation is an ancient practice that allows food to be grown where it might otherwise not thrive. Evidence shows that it was used as early as the sixth millennium B.C.E. in Mesopotamia, Egypt, and Persia, and the fifth millennium B.C.E. in South America. In the early twenty-first century, agriculture is still globally the greatest consumer of fresh water, though it varies widely by location. For example, the United Kingdom's abundant rainfall means that it requires almost no irrigation. Mexico and India, on the other hand, use it extensively. The green revolution of the twentieth century, which greatly increased the agricultural yield of many developing countries, relied in part on irrigation. One criticism was that the increased food production in these areas resulted in accelerated population growth that placed further burdens on scarce water resources. This criticism is supported by some statistics and mathematical models, which show that the demand for water grew at rates that exceeded population increases, raising per-capita water requirements.

Mathematicians and others who study ancient systems of irrigation in order to better understand them (and perhaps improve modern methods) have noted that some societies appear to have created and implemented complex and efficient water distribution methods without using mathematical methods for planning. Others have sought to build mathematical models of irrigation systems. The paddy field system used for growing rice generally requires the creation of intricate structures of terraces, canals, and reservoirs in order to ensure that all fields receive adequate water. It is believed to have been used as early as 4000–3500 B.C.E. in China and Korea. Researchers who have investigated mathematical models to describe a paddy field system have noted that it may not be possible to create a reliable model by including only variables based on physical measures such as amount of water available and rate of evaporation. A variable describing an ethic of cooperation among owners of the various fields, a factor that is difficult to quantify, was also required to ensure that water would be used fairly. For example, if owners on the upstream end of a water source took more than their fair shares, the owners farther downstream would not receive sufficient water for their crop, regardless of the values of some other variables.

Industry

Industry is the second largest category of global water use. Most industrial processes need water in some way, though some are more readily visible, such as hydropower generation of electricity and water extraction of minerals in mining. At the start of the twenty-first century, per capita water use is typically higher in industrialized nations than in developing countries, though this gap is closing. Some economists use the term "virtual water" to refer to the water that is used in the entire chain of manufacturing a product or growing an agriculture commodity. Similar to a carbon footprint, which is often used to quantify the quantity of greenhouse gasses emitted by a process, a water footprint represents the total amount of water used to create a good or service. Calculating water footprints provides an additional metric for assessing and comparing the environmental impact of competing products and services. For example, in 2010, the Water Footprint Network estimated that production of 1 kilogram of beef required about 16,000 liters of water, while one kilogram of rice required 3000 liters of water, and one liter of milk required 1000 liters of water.

Sanitation

The creation of sanitary systems of water supply and wastewater disposal or treatment is a major factor in the general improvement of public health from about the mid-nineteenth century onwards. Large cities, such as London, New York, and Boston, were among the first to establish municipal water supply systems. They were motivated in part by data collected by statisticians and others such as physician John Snow, who demonstrated via statistical methods that an 1854 cholera outbreak in London could be traced to the local water pump.

Mathematical methods may be used to model different aspects of supply systems. The fluid pressure necessary for water to flow through a system is affected by variables like gravity. Water stored in a rooftop tank will deliver water at a higher pressure to lower floors versus higher ones. Mathematical calculations show that a vertical foot of water exerts a pressure of 0.433 pounds per square inch (psi) at its bottom surface. The flow of water through the system is a function of the cross-sectional area of the pipe: $Q = A \times V$, where Q is the flow of water through the system, A is the cross-sectional area of the pipe, and V is the velocity of the water.

Municipal water systems tend to be quite complex, involving massive networks of storage tanks, pipes, pumps, and valves. Mathematical models are used to describe and manage these systems. Navier–Stokes equations, named for mathematicians Claude-Louis Navier and George Gabriel Stokes, are partial differential equations that describe fluid flow and velocity, while the Reynolds number, named for mathematician Osborne Reynolds, quantifies "laminar" (smooth) and turbulent fluid flow through a pipe. Contamination is an ever-present risk because of the natural physical deterioration of system components over time (such as corroded pipes) as well as the possibility of accidental or deliberate introduction of contaminants. Researchers are developing systems that can sense when a contaminant has been introduced into the water distribution system, allowing for rapid identification of the time and location of its introduction. For example, experiments done by the U.S. Environmental Protection Agency showed statistically that chlorine and total organic carbon, which are routinely monitored in municipal water systems, were sensitive and reliable predictors of contamination.

Further Reading

Cohen, Y. Koby. *Problems in Water Distribution: Solved, Explained, and Applied.* Boca Raton, FL: CRC Press, 2002.

Gates, James. *Applied Math for Water Distribution, Treatment, and Wastewater Operators.* Dubuque, IA: Kendall Hunt Publishing, 2010.

Sarah Boslaugh

Wheel

Category: Travel and Transportation.
Fields of Study: Algebra; Geometry.
Summary: Wheels help humans perform work and travel by providing a mechanical advantage.

Circles are present in many places in nature and mathematicians studied them long before the common use of the wheel. A wheel is traditionally a cylinder rotating around an axle. Together, a wheel and an axle form a simple machine that can change direction and magnitude of forces. Wheels are widely used in transportation as gears, as handles and knobs, and for converting the energy of water, animals, or people into work. The notion of curvature is of interest to many mathematicians, scientists, engineers, and others. In geometry, wheels are often modeled as circles or as concentric circles. In addition to standard circles or cylinders, mathematicians have explored the properties of wheels of other shapes along with varying surfaces. Aristotle's Wheel paradox, named for Aristotle of Stagira, is an interesting mathematical problem involving the paths traced by a wheel made of two concentric circles. It seems to imply that the circumferences of different sized circles are equal. This is one of many mathematical questions that arise from rotating concentric circles or exploring the curves generated by wheels.

History and Mechanical Advantage

Wheeled vehicles were invented about 6500 years ago, but they were not used widely until the rise of large, organized, road-building societies. This discrepancy between the discovery and its wide adoption, because of the lack of infrastructure, is frequent in science. Using wheels as levers to change the magnitude of force for applications like grinding grain was more widespread in many societies. The force advantage that a wheel provides is equal to the radius of the wheel divided by the radius of the axle. For example, a ship's capstan with the radius of eight feet and the axle radius of one foot multiplies the force of sailors using it by eight. This relationship is the reason that water wheels on small, weak streams that do not provide much force have to be larger than on fast-moving streams—a weak stream will not provide enough force to turn a small wheel. Rotating handles or knobs, grinders, drills, and old-fashioned water wells all use the wheel's mechanical advantage.

Geometry and Physics of Rolling: Work Smart, Not Hard

Rolling vehicles on wheels save work compared to dragging the same weight along the ground. Friction between the ground and a dragged object occurs along the length of the path. The work needed to overcome this friction is proportional to the friction coefficient, which depends on the surfaces of the object and the path. On smooth surfaces, such as ice, the friction coefficient is lower than on rough surfaces, such as rock. Work is also proportional to the weight of the object and the length of the path. When an object is rolled, its weight presses the axles to the wheels. Instead of the object-road friction, the force to overcome is now the axle-wheel friction, which is also proportional to the weight. When a wheel turns around, the vehicle travels the distance equal to the wheel's circumference. If the radius of the axle is one-tenth of the radius of the wheel, then the distance the axle slides within the wheel is one-tenth of the distance the vehicle travels and the required work is divided by 10. It is relatively easy to reduce axle-wheel friction many times by using smooth surfaces, oil, and ball bearings. Vehicles for heavier loads usually have more wheels to distribute the force of the load.

Reinventing the Wheel

Since wheels are essential to most human endeavors, there are many wheel-related sayings. "Reinventing the wheel" means "needlessly duplicating a well-known method." Ironically, wheels themselves are being constantly reinvented. For example, roller bearings first appeared in Leonardo da Vinci's drawings in the sixteenth century but were patented and used widely only in the nineteenth century. Magnetic bearings reduce axle-wheel friction to essentially zero and, therefore, promise huge increases in machine efficiency; their development started in 1980s. In the 1990s, mathematics and science museums began to feature bikes with square wheels that move smoothly over special surfaces consisting of "catenaries," which are hyperbolic shapes resembling hanging lengths of chains.

Further Reading

Farris, Frank. "Wheels on Wheels on Wheels—Surprising Symmetry." *Mathematics Magazine* 69, no. 3 (1996).

Goodstein, Madeline P. *Wheels! Science Projects With Bicycles, Skateboards, and Skates*. Berkeley Heights, NJ: Enslow Publishers, 2009.

Helfand, Jessica. *Reinventing the Wheel*. New York: Princeton Architectural Press, 2002.

Maria Droujkova

Windmills

Category: Architecture and Engineering.
Fields of Study: Algebra; Geometry; Measurement.
Summary: The amount of power that a windmill can harness can be determined mathematically according to its size and design.

For centuries, windmills have captured peoples' imaginations through their form, function, and romantic appeal. Immortalized by Miguel de Cervantes in his book *Don Quixote*, windmills have transformed over the years from broad, short structures with an even number of sails to tall, sleek, three-sailed structures equipped with turbines for capturing energy from the wind. Windmills utilize natural power sources to perform a variety of functions, including energy production and food processing. Wind-driven prayer wheels have been used since the fourth century in Tibet and China. Historians believe that people in ancient Persia built the first practical windmills for both grinding grain and pumping water. From there, they spread through the Middle East and parts of Asia, as well as to India. They can be documented in Europe by the twelfth century. Wind turbines developed primarily in the twentieth century. Mathematics has been important for both the design of windmills and in calculating and modeling their output. Interestingly, English mathematician and physicist George Green was also a miller, and he is believed to have done much of his mathematics work in his windmill.

Designs

Windmills have had a wide variety of designs and appearances. Some of the earliest windmills rotated along a vertical axis, with the main rotor placed vertically in relation to the ground and giving a look similar to a helicopter. Some modern wind turbines have retained this engineering design in areas where wind direction is variable. This design is advantageous because vertical-axis windmills have an axis of rotation perpendicular to the ground, so the sails react similarly to all wind directions. On the other hand, horizontal-axis windmills have an axis of rotation that is parallel to the ground, resembling the more common image of a windmill such as that found in *Don Quixote*. The structure of horizontal-axis windmills gives the advantage of allowing their potential work to be maximized with respect to a specific wind direction. It is important to place a horizontal-axis windmill in line with the prevailing wind.

Windmills have traditionally been designed symmetrically, including an even number of sails. Historically, workers would place food and other substances in special locations inside the windmill to be ground by stones or other clashing materials. The grinding materials were sometimes connected to a system of gears and pulleys to increase the power beyond the mere rotation of the sails. Most modern wind turbines continue to have a sleek, symmetric design but have three sails. The insides of these turbines are devoted mostly to the attainment of electric power.

Number of Blades

The number of blades on a windmill is in direct correlation to the power generated, although the coefficient is quite small. The amount of power generated increases nearly linearly with each additional blade but the increase in power beyond just two or three blades is quite small for modern wind turbines. Physicists have determined that the power generated by a wind turbine is proportional to the cube of the wind speed and can be found algebraically by

$$P = EA\frac{1}{2}dv^3$$

where E is the power efficiency of the rotor, A is the swept area, d is air density, and v is wind speed. The swept area relates to the circle created by a rotation of a sail, calculated by

$$A = \frac{1}{2}\pi l^2.$$

where l is the length of the sail. The theoretical maximum of E, known as the "Betz limit," is 0.59. The Betz limit is named for Albert Betz, a German physicist who was also interested in wind power. However, this theoretical value is reduced significantly when common physical constraints, including friction and drag on the rotors, are considered. One can calculate the maximum power produced by a windmill algebraically as

$$P_{max} = \frac{8}{27}A\frac{1}{2}dv^3.$$

It is difficult to put tight parameters on the variables that determine the amount of power produced by a wind turbine. However, a good estimate of the production of power for a 10-foot diameter sail in 12 miles per hour average winds is 2300 kilowatts of power. In a wind farm, several turbines are interconnected by a power collection system and communications network to pool their output and connect to a power grid. Probabilistic mathematical models are used to estimate and describe the output of networks of wind turbines.

Further Reading

Betz, A. *Introduction to the Theory of Flow Machines.* Translated by D. G. Randall. Oxford, England: Pergamon Press, 1966.

Brooks, L. *Windmills.* New York: Metro Books, 1989.

Gipe, Paul. *Wind Power, Revised Edition: Renewable Energy for Home, Farm, and Business.* New York: Chelsea Green Publishing, 2004.

Gorban, A. N., A. M. Gorlov, and V. M. Silantyev. "Limits of the Turbine Efficiency for Free Fluid Flow." *Journal of Energy Resources Technology,* 123, no. 4 (2001).

David Slavit
Gisela Ernst-Slavit

Wireless Communication

Wireless communication has become ubiquitous in the twenty-first century. Consider all of the aspects of one's life that are impacted by wireless communications, including text messaging and voice calls over a cellular network, and e-mail and Web surfing over a wireless Internet connection. Wireless communication consists of encoding information onto radio waves and passing them through the atmosphere—not unlike how an amplitude modulation (AM) or frequency modulation (FM) radio signal is sent and received. Wireless communication would not be possible without mathematics, and mathematicians contribute in many ways to creating, sustaining, and studying wireless processes and technologies.

Information theory plays a central role in wireless communications; its origins are attributed to mathematician Claude Shannon in the mid-twentieth century. Sergio Verdu, who is cited as a world-renown researcher in wireless communications noted, "Claude Shannon was the archetypical seamless combination of mathematician and engineer. . . . Shannon's theory has been instrumental in anything that has to do with modems, wireless communications, multi-antenna and so on."

Many other theoretical and applied mathematical methods have also been fundamental in wireless communication. For example, methods like stochastic calculus, stochastic modeling, control theory, graph theory, game theory, signal processing, wavelets, simulation and optimization, and multivariate statistical analysis have been used to develop communication networks, quantify or predict performance characteristics like network traffic, and to create protocols for signal transmission, encryption, and compression. Some mathematical models have been used by developers to quantify and compare wired versus wireless communication systems.

Mathematicians and engineers working in wireless communications must consider the properties of the waves and how the information is encoded. Information, whether an e-mail, telephone, video, or other data, is encoded onto the sinusoidal waveform by combining changes in frequency, amplitude, and phase. This encoding is accomplished by modifying various properties of a periodic sinusoidal function—the carrier wave—to embed information or message wave on the carrier. Figure 1 shows a simple example for the case of AM. The height or amplitude of the carrier wave is modified to represent or information or modulating wave.

Researchers also consider the variety of factors that can affect the strength and quality of the signal. A communications engineer or technician is most often concerned with behaviors that will affect the propagation of the radio wave through the air. These include absorption, attenuation, diffraction, free space path loss, gain, reflection, refraction, and scattering. A combination of these factors will impact the signal quality and determine the likelihood of a successful transmission.

One common number associated with a wireless signal is the frequency. Frequency is a measure of how many cycles occur for a given time period. A signal cycle occurs every time a waveform repeats. Frequency is measured in cycles per second, which are also called "hertz" (Hz) after German physicist Heinrich Hertz. A waveform that repeats once every second has a frequency of 1 hertz. Waves used in communications are at much higher frequencies, so some prefixes must be used to measure radio frequencies. The wireless networks used for laptops and smartphones at the beginning of the twenty-first century often operate at the 2.4 GHz and 5 GHz frequencies of the spectrum. AM and FM radio are in the kHz or MHz frequencies, while

Figure 1. Amplitude modulation (AM).

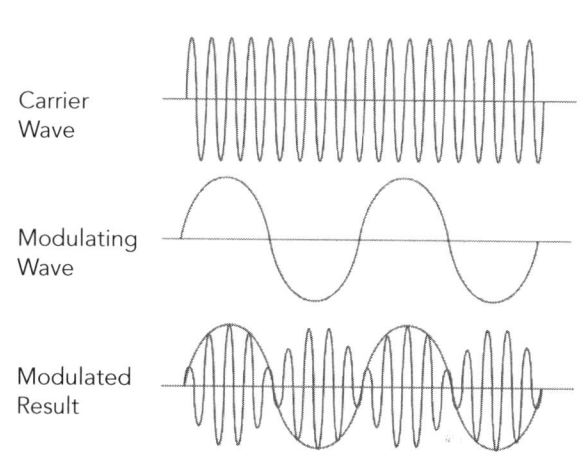

satellites operate at very high frequencies—often in the hundreds of GHz.

Further Reading

Agrawal, Prathima, Daniel Andrews, Philip Fleming, George Yin, and Lisa Zhang. "Wireless Communications." In *IMA Volumes in Mathematics and its Applications Series*. Vol. 143. New York: Springer, 2007.

Boche, Holger, and Andreas Eisenblatter. "Mathematics in Wireless Communications." In *Production Factor Mathematics*. Edited by Martin Grotschel, Klaus Lucas, and Volker Mehrmann. Berlin: Springer, 2010.

Leong, Y. K. "Mathematical Conversations—Sergio Verdu: Wireless Communciations, at the Shannon Limit." *National University of Singapore Newsletter of the Institute for Mathematical Sciences* 11 (September 2007). http://www.princeton.edu/~verdu/singapore.pdf.

Michael Qaissaunee

Resource Guide

Books

Aaboe, Asger. *Episodes From the Early History of Mathematics*. Washington, DC: Mathematical Association of America, 1975.

Adrian, Yeo. *The Pleasures of Pi and Other Interesting Numbers*. Singapore: World Scientific Publishing, 2006.

Agresti, A. *Categorical Data Analysis*. Hoboken, NJ: Wiley, 2002.

Aho, A. V., J. E. Hopcrotf, and J. D. Ullman. *The Design and Analysis of Computer Algorithms*. Reading, MA: Addison-Wesley, 1976.

Albert, Jim, and Jay Bennett. *Curve Ball: Baseball, Statistics, and the Role of Chance in the Game*. New York: Springer-Verlag, 2001.

Ascher, Marcia. *Mathematics Is Everywhere: An Exploration of Ideas Across Cultures*. Princeton, NJ: Princeton University Press, 2002.

Ball, W. W. Rouse. *A Short Account of the History of Mathematics*. New York: Sterling Publishing Company, 2001.

Barnett, Raymond, Michael Ziegler, and Karl Byleen. *Calculus for Business, Economics, Life Science, and Social Science*. Upper Saddle River, NJ: Prentice-Hall, 2005.

Baumohl, Bernard. *The Secrets of Economic Indicators: Hidden Clues to Future Economic Trends and Investment Opportunities*. 2nd ed. Upper Saddle River, NJ: Pearson Education, 2008.

Beckmann, Petr. *A History of π (Pi)*. New York: Barnes & Noble, 1971.

Behrends, Ehrhard. *Five-Minute Mathematics*. Providence, RI: American Mathematical Society, 2008.

Bell, Eric Temple. *Men of Mathematics*. New York: Simon & Schuster, 1937.

Bennett, Jay, and James Cochran. *Anthology of Statistics in Sports*. Philadelphia, PA: Society for Industrial and Applied Mathematics, 2005.

Berggren, Lennart, Jon Borwein, and Peter Borwein. *Pi: A Source Book*. New York: Springer-Verlag, 1997.

Berlekamp, Elwyn R., John H. Conway, and Richard K. Guy. *Winning Ways for Your Mathematical Plays*. Natick, MA: AK Peters, 2001.

Blackwell, William. *Geometry in Architecture*. Hoboken, NJ: Wiley, 1984.

Blatner, David. *The Joy of π*. New York: Walker & Co., 1997.

Blue, Ron, and Jeremy White. *The New Master Your Money: A Step-by-Step Plan for Gaining and Enjoying Financial Freedom*. Chicago: Moody, 2004.

Blum, Raymond. *Mathemagic*. New York: Sterling Publishing, 1992.

Bodie, Zvi, Alex Kane, and Alan Marcus. *Investments*. Chicago, IL: McGraw-Hill/Irwin, 2008.

Borwein, Jonathan, and Peter Borwein. *A Dictionary of Real Numbers*. Pacific Grove, CA: Brooks/Cole Publishing Co., 1990.

Boyer, C. B. *A History of Mathematics*. Hoboken, NJ: Wiley, 1968.

Boyer, C. B. *The History of the Calculus and Its Conceptual Development*. New York: Dover Publications, 1949.

Brealey, Richard A., Stewart C. Myers, and Franklin Allen. *Principles of Corporate Finance*. 9th ed. New York: McGraw-Hill, 2008.

Bressoud, David. *The Queen of the Sciences: A History of Mathematics*. Chantilly, VA: The

Teaching Company, 2008.

Broverman, Samuel A. *Mathematics of Investment and Credit*. Winsted, CT: ACTEX Publications, 2008.

Burkett, Larry, and Brenda Armstrong. *Making Ends Meet: Budgeting Made Easy*. Gainesville, GA: Crown Financial Ministries, 2004.

Burton, David M. *The History of Mathematics: An Introduction*. New York: McGraw-Hill, 2005.

Calinger, Ronald. *A Contextual History of Mathematics*. Upper Saddle River, NJ: Prentice-Hall, 1999.

Clagett, Marshall. *Archimedes in the Middle Ages*. Madison: University of Wisconsin Press, 1964.

Closs, Michael. *A Survey of Mathematics Development in the New World*. Ottawa: University of Ottawa, 1977.

Closs, Michael, ed. *Native American Mathematics*. Austin: University of Texas Press, 1986.

Coe, Michael D. *Breaking the Maya Code*. New York: Thames and Hudson, 1992.

Copeland, Thomas E., J. Fred Weston, and Kuldeep Shastri. *Financial Theory and Corporate Policy*. 4th ed. Upper Saddle River, NJ: Pearson Education, 2005.

Cullen, Christopher. *Astronomy and Mathematics in Ancient China: The Zhou Bi Suan Jing*. Cambridge, England: Cambridge University Press, 1996.

Cuomo, Serafina. *Ancient Mathematics*. London: Routledge, 2001.

Davenport, Harold. *The Higher Arithmetic: An Introduction to the Theory of Numbers*. Cambridge, England: Cambridge University Press, 1999.

Davis, Morton D. *The Math of Money: Making Mathematical Sense of Your Personal Finances*. New York: Copernicus, 2001.

De Mestre, Neville. *The Mathematics of Projectiles in Sport*. Cambridge, England: Cambridge University Press, 1990.

Devlin, Keith. *The Math Gene: How Mathematical Thinking Evolved and Why Numbers Are Like Gossip*. New York: Basic Books, 2001.

———. *The Unfinished Game: Pascal, Fermat, and the Seventeenth-Century Letter That Made the World Modern*. New York: Basic Books, 2008.

Drobot, Stefan. *Real Numbers*. Upper Saddle River, NJ: Prentice-Hall, 1964.

Dudley, Underwood. *Numerology or What Pythagoras Wrought*. Washington, DC: Mathematical Association of America, 1997.

Eastway, Rob, and John Haigh. *Beating the Odds: The Hidden Mathematics of Sport*. London: Robson Books, 2007.

Eglash, Ron. *African Fractals: Modern Computing and Indigenous Design*. New Brunswick, NJ: Rutgers University Press, 1999.

Eves, Howard. *An Introduction to the History of Mathematics*. New York: Saunders College Publishing, 1990.

Flegg, G. *Numbers: Their History and Meaning*. New York: Schocken Books, 1983.

Friberg, Jöran. *Unexpected Links Between Egyptian and Babylonian Mathematics*. Singapore: World Scientific Publishing Co., 2005.

Friedman, Arthur. *World of Sports Statistics: How the Fans and Professionals Record, Compile, and Use Information*. New York: Athenaeum, 1978.

Fries, Christian. *Mathematical Finance: Theory, Modeling, Implementation*. Hoboken, NJ: Wiley, 2007.

Frumkin, Norman. *Guide to Economic Indicators*. Armonk, NY: M. E Sharpe, 2000.

Gamow, George. *One, Two, Three... Infinity*. New York: Viking Press, 1947.

Gardner, David, and Tom Gardner. *The Motley Fool Personal Finance Workbook: A Foolproof Guide to Organizing Your Cash and Building Wealth*. New York: Fireside Books, 2003.

Gardner, Martin. *Mathematics, Magic and Mystery*. New York: Dover, 1956.

Gay, Timothy. *The Physics of Football*. New York: HarperCollins, 2005.

Gerdes, Paulus. *Geometry From Africa: Mathematical and Educational Explorations*. Washington, DC: Mathematical Association of America, 1999.

Gillings, R. J. *Mathematics in the Time of the Pharaohs*. New York: Dover Publications, 1982.

Gutstein, Eric, and Bob Peterson, eds. *Rethinking Mathematics: Teaching Social Justice by the Numbers*. Milwaukee, WI: Rethinking Schools, 2005.

Hadamard, Jacques. *A Mathematician's Mind*. Princeton, NJ: Princeton University Press, 1996.

Hardy, G. H. *A Mathematician's Apology*. Cambridge, England: Cambridge University Press, 1941.

Henry, Granville C. *Logos: Mathematics and Christian Theology*. Lewisburg, PA: Bucknell University Press, 1976.

Hersh, Rueben. *What Is Mathematics, Really?* New York: Oxford University Press, 1997.

Resource Guide 137

Hoyle, Joe Ben, Thomas F. Schaefer, and Timothy S. Doupnik. *Fundamentals of Advanced Accounting*. New York: McGraw-Hill, 2010.

Kalbfleisch, John D., and Ross L. Prentice. *The Statistical Analysis of Failure Time Data*. Hoboken, NJ: Wiley, 2002.

Katz, Victor J., ed. *Mathematics of Egypt, Mesopotamia, China, India, and Islam: A Sourcebook*. Princeton, NJ: Princeton University Press, 2007.

Kellison, Stephen G. *Theory of Interest*. New York: McGraw-Hill, 2009.

Kimmel, Paul D., Jerry J. Weygandt, and Donald E. Keiso. *Financial Accounting: Tools for Business Decision Making*. Hoboken, NJ: Wiley, 2009.

King, Jerry. *The Art of Mathematics*. New York: Plenum Press, 1992.

Klein, John P., and Melvin L. Moeschberger. *Survival Analysis: Techniques for Censored and Truncated Data*. New York: Springer-Verlag, 1997.

Kline, M., *Mathematical Thought From Ancient to Modern Times*. New York: Oxford University Press, 1972.

Koetsier, T., and L. Bergmans, eds. *Mathematics and the Divine: A Historical Study*. Amsterdam: Elsevier, 2005.

Longe, Bob. *The Magical Math Book*. New York: Sterling Publishing, 1997.

Martzloff, Jean-Claude. *A History of Chinese Mathematics*. New York: Springer-Verlag, 1987.

Moses, Robert P., and Charles E. Cobb, Jr. *Radical Equations: Civil Rights From Mississippi to the Algebra Project*. Boston: Beacon Press, 2001.

Mullis, Darrell, and Judith Handler Orloff. *The Accounting Game: Basic Accounting Fresh From the Lemonade Stand*. Naperville, IL: Sourcebooks, 2008.

Nahin, Paul J. *Dr. Euler's Fabulous Formula*. Princeton, NJ: Princeton University Press, 2006.

Nasar, Sylvia. *A Beautiful Mind: The Life of Mathematical Genius and Nobel Laureate John Nash*. New York: Simon & Schuster, 2001.

Oliver, Dean. *Basketball on Paper: Rules and Tools for Performance Analysis*. Washington, DC: Brassey's, 2004.

Pullan, J. M. *The History of the Abacus*. New York: F. A. Praeger, 1969.

Rafiquzzaman, M. *Fundamentals of Digital Logic and Microcomputer Design*. Hoboken, NJ: Wiley, 2005.

Rudin, W. *Principles of Mathematical Analysis*. New York: McGraw-Hill, 1953.

Salem, Lionel, Frédéric Testard, and Coralie Salem. *The Most Beautiful Mathematical Formulas*. Hoboken, NJ: Wiley, 1992.

Schwarz, Alan. *The Numbers Game: Baseball's Lifelong Fascination with Statistics*. New York: St. Martin's Press, 2004.

Smith, D. E. *History of Mathematics*. Vol. 2. New York: Dover Publications, 1958.

Solow, Daniel. *How to Read and Do Proofs: An Introduction to Mathematical Thought Process*. Hoboken, NJ: Wiley, 1982.

Steen, Lynn A. *On the Shoulders of Giants: New Approaches to Numeracy*. Washington, DC: National Academy Press, 1990.

Sterrett, Andrew. *101 Careers in Mathematics*. Washington, DC: The Mathematical Association of America, 1996.

Suzuki, Jeff. *A History of Mathematics*. Upper Saddle River, NJ: Prentice Hall, 2002.

Taylor, Alan D. *Mathematics and Politics: Strategy, Voting Power, and Proof*. New York: Springer-Verlag, 1995.

van der Waerden, B. L. *Geometry and Algebra in Ancient Civilizations*. Berlin: Springer, 1983.

Venema, G.A. *The Foundations of Geometry*. Upper Saddle River, NJ: Pearson Prentice Hall, 2006.

Weygandt, Jerry J., Paul D. Kimmel, and Donald E. Keiso. *Managerial Accounting: Tools for Business Decision Making*. Hoboken, NJ: Wiley, 2008.

Winkler, Peter. *Mathematical Puzzles: A Connoisseur's Collection*. Natick, MA: AK Peters, 2004.

Wright, Tommy, and Joyce Farmer. *A Bibliography of Selected Statistical Methods and Development Related to Census 2000*. Washington, DC: U.S. Bureau of the Census, 2000.

Yeldham, F. A. *The Teaching of Arithmetic Through Four Hundred Years (1535–1935)*. London: G. G. Harrap & Company, 1935.

Yong, L. L., and A. T. Se. *Fleeting Footsteps*. Singapore: Word Scientific Publications, 2004.

Zaslavsky, Claudia. *Africa Counts: Number and Pattern in African Culture*. Chicago: Lawrence Hill Books, 1999.

Zill, D. G. *Calculus with Analytic Geometry*. Boston: Prindle, Weber & Schmidt, 1985.

Journals and Magazines

The AMATYC Review
The American Mathematical Monthly
Association for Women in Mathematics Newsletter

Biometrics
Chance
The College Mathematics Journal
Experimental Mathematics
The Fibonacci Quarterly
Historia Mathematica
IMU-Net
Involve
Journal of Humanistic Mathematics
Journal of Integer Sequences
Journal of Recreational Mathematics
Journal of Statistics Education
Loci
MAA FOCUS
Math Horizons
Mathematics Magazine
Mathematics Teacher
NAM Newsletter
Notices of the American Mathematics Society
The Pentagon
Pi Mu Epsilon Journal
Plus Magazine
PRIMUS
Rose-Hulman Undergraduate Mathematics Journal
SIAM Review
Scholastic Math
Significance
Teaching Children Mathematics
Undergraduate Mathematics and Its Applications

Internet
American Institute of Mathematics
    www.aimath.org
The Algebra Project
    www.algebra.org
AMATYC
    www.amatyc.org
American Mathematical Society
    www.ams.org

American Statistical Association
    www.amstat.org
Association for Women in Mathematics
    www.awm-math.org
CryptoKids
    www.nsa.gov/kids
Datamath Calculator Museum
    www.datamath.org
Illuminations
    illuminations.nctm.org
MacTutor History of Mathematics
    www-history.mcs.st-and.ac.uk
Mathematical Fiction
    http://kasmana.people.cofc.edu/MATHFICT
Math for America
    www.mathforamerica.org
Math Forum
    www.mathforum.com
Math Fun Facts!
    www.math.hmc.edu/funfacts
MathDL
    mathdl.maa.org/mathDL
Mathematical Association of America
    www.maa.org
Mathematical Science Research Institute
    www.msri.org
The Museum of Mathematics
    www.momath.org
National Association of Mathematicians
    www.nam-math.org
National Council of Teachers of Mathematics
    www.nctm.org
RadicalMath
    www.radicalmath.org
Society for Industrial and Applied Mathematics
    www.siam.org
We Use Math
    www.weusemath.org
Wolfram MathWorld
    www.mathworld.wolfram.com

Index

Text and page numbers in **boldface** refer to main topics.

3-D graphing calculators, 17
3-D televisions, 109

Abstract Linking Electronically (ABLE), 16
Adleman, Leonard, 31
Adrain, Robert, 33
Advanced Placement (AP) Exams, 15
Advanced Placement Calculus (AP Calculus), 17
Advanced Research Projects Agency Network (ARPANET), 63, **64**
Advances in Mathematics of Communications (journal), 37
aerodynamics, 1, 4
aircraft design, 1–4
    aircraft carriers, 2
    complex analysis and, 1
    Joukowski airfoil, 1
    nature-ispired algorithms, 1
    sonic booms, 2
airplanes/flight, 4–6
    flight speed, 5
    George Cayley and, 4
    lift and thrust, 5
    mathematical history of, 4
    principles of flight, 5
algorithms
    Huffman, 52
    nature-inspired, 1
    traveling salesman problem and, 120
Alhambra Palace (Spain), 23
alternating current (AC), 47
Alvin, 39

Amazon, 40, **67**
Amdahl, Gene, 85
American City Planning Institute, 27
American Journal of Physics, 122
American Society for Communication of Mathematics, 34
Ameritech, 24
Ampère, André-Marie, 70
amperes, 70
amplitude modulation (AM), 90, **132**
animation and CGI, 6–8
    early devices, 6
    mathematics and, 7
    principles of, 7
Apple Computer, Inc., 67, 86
Appleton, Edward, 90
Archimedes
    contributions, 69
    Equilibrium of Planes, 69
    levers and, 69
ard disk drives (HDDs), 45
arenas, sports, 8–9
Aristotle, 41, **130**
Aristotle's Wheel Paradox, 130
Armstrong, Lance, 13
ARPANET, 97, **98**
artificial neural networks, 78, **80**
ArXiv.org e-print archive, 33
A Standard City Planning Enabling Act (U.S. Dept. of Commerce), 27
A Standard State Zoning Enabling Act (U.S. Dept. of Commerce), 27
astronomy
    lunar calendars, 18

AT&T, 24
audio processing, 76
automata, 124
automobiles
    city planning and, 25, **27**
    design of, 50
    highway design, 60, 61
    Segways versus, 96
    smart car, 103
    thermostats, 111
    traffic modeling, 117
auto racing, 9–11
    car design, 10
    overview, 9
    race strategy, 11
    race track design, 10
    technology and safety, 11

Babbage, Charles
    mechanical computers and, 86, **91**
Bain, Alexander, 52
Bakewell, Frederick, 52
Baran, Paul, 64
Bardeen, John, 86
Barnes & Noble Nook, 40
bathyspheres, 39
Bayesian decision theory, 92, **108**
Bayes, Thomas, 79
bearings, 130
Beaufort cipher, 31
Beaufort, Francis, 31
Becher, Johann, 124
Bell Telephone Laboratories, 29, **43**
Bell X-1 rocket-propelled airplane, 6

139

Berners-Lee, Tim, 97
Bernoulli, Daniel
    fluid force studies and, 4
Bernoulli's principle, 4
Betz, Albert, 132
bicycles, 11–13
Binary Automatic Computer (BINAC), 86
bin-packing problem, 99
biomimicry, 59
bitmap graphics, 43
BITNET, 64
black holes, 107
blocks and tackles, 88
Blondel, Vincent, 24
Boldyrev, Dmitry, 77
Boolean algebra (Boolean logic), 30
Booth, Charles, 26
Bourne, William, 39
Boussinesq, Joseph, 19
Boyle, Willard, 43
brain
    optical illusions and, 82
Bramer, Benjamin, 22
Brandenburg, Karl-Heinz, 77
Brattain, Walter, 86
bridges, 13–14
Brisson, Barnabé, 19
Brodetsky, Selig, 4
Bureau International de l'Heure (The International Time Bureau), 30
Burj Khalifa, 101
business, economics, and marketing
    credit, 37
    Internet and, 67
Buys-Ballot, Christophorus, 118

Caesar, Julius, 18, 30
Calculated Industries (CI), 16
calculators in society, 14–19
    early history of, 15
    graphing calculators, 17
    special-purpose, 16
Calder, Alexander, 70
calendars, 17–19
    Julian and Gregorian, 18
    lunar, 18
    Mayan, 19
    solar, 19
Cal-Tech (calculator), 15
canals, 19–20
Capek, Karel, 91

carbon footprints
    calculation of, 58
Cardan grill, 30
careers
    in communications, 33
    employers of mathematicians, 32
car-following traffic modelling, 116
carpentry, 20–22
cartography, 71
Casio, 17
castles, 22–24
Cathode-Ray Tube Amusement Device, 126
CAT scans, 43
Cayley, George, 4
CCD chips, 41, 43
CD drives, 86
CDs, 46
Cedar Point Amusement Park (Ohio), 94
cell phone networks, 24–25
cell phones, 17
cell phone towers, 24
central processing units (CPUs), 84, 86
centripetal force, 94
Cerf, Vinton, 98
CERN httpd (W3C httpd), 97
Chamberlain, Nira, 3
Chartier, Timothy, 7
Chebyshev, Pafnuty, 91
Chiao Wei-Yo, 20
cholera outbreak, 25, 129
Church, Alonso, 86
cipher-text, 30
city planning, 25–28, 26, 57
classic mathematical problems
    bin-packing problem, 99
    traveling salesman problem (TSP), 99, 120
clock arithmetic, 28
clocks, 28–30
CMOS chips, 41
Coandă effect, 5
Coandă, Henri, 5
Cobb, Paul, 36
code talking, 90
coding and encryption, 30–32
    code talking, 90
    credit card encryption, 37
    fax machines, 52, 52–106
cognitive psychology, 83
coil springs, 73

Cold War
    arms race, 64
communication in society, 32–37
    mathematical applications/technologies and, 36
    mediums for, 33
    proofs and, 35
complex analysis, 1
computer aided design (CAD), 21, 50, 52
computerized axial tomography (CAT scanning), 43
computers
    early history of, 64
    hacking, 37
    mechanical, 91
Construction Master Pro software, 17
contaminants, 129
continuum modeling, 117
Cooper, Paul, 122
coordinate geometry, 51
Coriolis effect, 115
Coulomb, Charles-Augustin de, 47
Coulomb's Law, 47
credit cards
    encryption, 37
Crispin, Mark, 64
CRT televisions, 110
cryptology, 36
Cummings, Alexander, 115
Cybernetics (Weiner), 92
cyclic redundancy check (CRC), 46
cycloid, 29

Dalgarno, George, 124
dams, 37–38
Darcy-Weisbach equation, 4
data compression, 77
data rot, 46
Davies, Donald, 64
De Arte Combinatoria (Leibniz), 123
deep submergence vehicles, 39–40
DeRose, Tony, 7, 76
design principles, 68
Devol, George, 92
differential GPS (DGPS), 56
digital book readers, 40–41
digital cameras, 41–43
digital images, 43–44
Digital Opportunity Index (DOI), 87
digital storage, 45–46
dikes, 19

direct current (DC), 47
discrete finite state automata (DFA), 124
Domain Name System (DNS) servers, 65
Doppler radar, 118
dot-com bubble, 67
DVD drives, 86
DVDs, 46
DVR devices, 81

Eastman, George, 41
Eastman Kodak Company, 41
Eberlein, Ernst, 77
Eckert, John Presper, 86
Edison, Thomas, 70, 110
Educate to Innovate campaign, 126
Edward I (King), 23
Eiffel, Gustave, 101
Eiffel Tower, 100
Einstein, Albert
    mathematics/reality link and, 123
Electrical Numerical Integrator and Calculator (ENIAC), 86
electricity, 46–48
electrodynamics, 70
electromagnetic radiation (EMR), 74, 89
electronic ink, 40
electronic passwords, 64
elevators, 48–50
Energy Policy Act, 115
Engelberger, Joseph, 92
engineering design, 50–52
Enigma code, 30
Equilibrium of Planes (Archimedes), 69
equilibrium theory, 66
Erdös, Paul, 66
Erdös-Rényi graphs, 66
Erie Canal, 20
Error-Correcting Code (ECC) memory, 98
Euclid of Alexandria
    pinhole camera and, 41
Euler, Leonhard
    Seven Bridges of Konigsberg and, 14
Eupalinian aqueduct, 121
Eupalinos of Megara, 121
European Organization for Nuclear Research (CERN), 97

Facebook, 106
fax machines, 52, 52–54

Federal Aid Highway Acts, 60
Federal Communications Commission, 24
feng shui, 68
Fibonacci sequence, 9
FidoNet, 64
Fignon, Laurent, 13
finite state machines, 124
First-fit algorithm, 99
first-generation (1G) cell technology, 24
Fischer, Carl, 15, 16
flash memory, 46
Fletcher, Thomas, 106
Flynn, Morris, 62
FONE F3, 40
Fourier transforms, 78, 87
fourth-generation (4G) cell technology, 25
fractals
    coastlines and, 71
    in village design, 68
Frankenstein (Shelley), 93
Franklin, Benjamin, 100
fraud detection
    in communication technologies, 36
Freeman, Greydon, 64
frequency modulation (FM), 90, 132
Fuchs, Ira, 64
fuel consumption, 54–55
Fuller, Buckminster "Bucky", 9
fx-7000G, 17

Galerkin, Boris, 38
Galileo (Galileo Galilei)
    pendulum clocks and, 29
    Two New Sciences, 69
Gates, Bill, 86
Gauss, Carl F.
    parallel processing and, 84
    thermostats and, 112
Gaussian thermostat, 112
Gauss, Karl Friedrich, 33
geographic information system (GIS), 71
geometry and geometry education
    of castle defense, 22
Geometry Forum, 34
German ciphers, 30
gestalt psychology, 84
g-forces, 94
Gini, Corrado, 87
Glauert, Hermann, 6

gnomonics, 28, 68
Golden Ratio, 110
Goldsmith, Thomas, 126
Google, 65, 67
GPS, 55–57
    smart cars and, 103
    trilateration and, 55, 56
graphing calculators, 17
graph theory, 66, 79
green design, 57–59
Green, George, 19, 131
Gregory XIII (pope), 18
Grill, Bernhard, 77
Gross, Mark, 93
ground resonance effect, 60
Guo Shoujing (Kuo Shou-ching), 20

hacking, computer, 37
harmonics, 60
Harrington, John, 115
Harrison, John, 29
Haussmann, Baron, 26
Heaviside Layer, 90
Heaviside, Oliver, 90
helicopters, 59–60
Helmholtz, Herman, 83
Henlein, Peter, 28
Hermann grid, 82
Heron of Alexandria (Hero), 121
Herschel, John, 41
Hertz, Heinrich, 47, 133
Hertz (Hz), 133
Hewlett, Bill, 15
Hewlett-Packard (HP), 15, 16, 17
high occupancy toll (HOT) lanes, 62
high occupancy vehicle (HOV) lane management, 61–63
highways, 60–61, 62, 116, 117
Holland, Clifford, 121
Home Insurance Building, 101
Hopfield neural networks, 79
Horner, William George, 6
horologium, 28
Horton, Joseph Warren, 29
Howard, Ebenezer, 27
"How Long Is the Coast of Britain? Statistical Self-Similarity and Fractional Dimension", 72
Huffman coding, 52, 77
Huffman, David, 52, 78
Hutton, Charles, 13
Huygens, Christian, 29

Index

hydraulics, 19, **48**
hydrodynamic modeling, 116
hydroelectric power, 37
hypersonic aircraft, 6

IBM Corporation, 86
ID3 data blocks, 78
Illuminations Web site, 34
illusions, optical, 82, **83**
imaging technologies, 111
Incan and Mayan mathematics
    calendars, 19
Indianapolis 500, 9
information theory, 30, **132**
innerspring mattresses, 73
Intel Corporation, 86
International Bureau of Weights and
    Measures (BIPM), 30
International Earth Rotation and Refer-
    ence Systems Service (IERS), 30
International Space Station, 92
International Telecommunications
    Union, 52
internet, 63–67
    economics and, 67
    history of, 63
    mathematical problems and, 64
    mathematical sciences codevelop-
        ment and, 63
    networks and, 65
Internet
    MP3 players and, 77
Internet Assigned Numbers Authority, 65
Internet Message Access Protocol
    (IMAP), 64
Internet Protocol (IP), 98
interprocess communication (IPC), 98
irrigation, **128**, 129
iTunes, 77

Jacquard, Joseph, 91
Jacquard loom, 91
jamitons, 62
Jansky, Karl Guthe, 90
Japan, 124
Jenney, William Le Baron, 101
Jennings, Thomas, 64
Joukowski airfoil, 1

Kahn, Robert E., 98
Kakutani, Shizuo, 101
Kasimov, Aslan, 62

Kennelly, Arthur Edwin, 90
Kennelly-Heaviside layer, 89
Kenschaft, Patricia, 89
Kilby, Jack, 15
kilowatts (kWh), 47
Kindle (Amazon), 40
Kircher, Athanasius, 124
Klein 4-group, 74
Klein, Felix
    Klein 4-group, 74
Kleinrock, Leonard, 64
knots, 68
Knowlton, Nancy, 102
Knuth, Donald, 34
Koopmans, Tjalling, 100
Kramer, Briton, 77
Kuo Shou-ching (Guo Shoujing), 20
Kurten, Bernd, 77

labyrinths, 68
Lake Pontchartrain Causeway (Louisi-
    ana), 14
Lambda calculus, 86
landscape design, 67–69
LaTeX, 34
Lavrentev, Mikhail, 19
Lax, Peter, 4
LCD televisions, 109
Leadership in Energy and Environmen-
    tal Design (LEED), 58
Lego Group, 93
Leibniz, Gottfried Wilhelm
    calculus and, 123
    De Arte Combinatoria, 123
    symbolic language and, 123
LeMond, Greg, 13
Leonardo da Vinci
    flight studies, 4
levers, 69–70
Librié, 40
light bulbs, 70–71
Lighthill-Whitham-Richards (LWR)
    traffic theory, 117
linkages, 69
Lockheed Corporation, 6
locomotives, 118
Lorenz, Max, 87
Lucasfilm LTD, 7
luminous efficacy, 71
Luotonen, Ari, 97

Mach bands, 82

Mach, Ernst, 2
Mach Number (M), 2, 6
Macintosh computers, 86
MacTutor History of Mathematics
    Archive, 34
magnetic disk drives, 86
magnetic tunnels, 87
Mandelbrot, Benoît, 72
Manin, Yu, 36
Mann, Estle Ray, 126
Maple (software), 86
mapping coastlines, 71–73
Marconi, Guglielmo, 89
Marianas Trench, 39
Markham, Beryl, 6
Marrison, Warren, 29
Martin, David, 102
Mathcad software, 9
"math castle", 23
Mathematical Association of America
    (MAA), 34
mathematical engines, 91
Mathematical Markup Language
    (MathML), 34
mathematical modeling
    of auditory processing, 78
    cycling equipment and, 13
    for helicopter flight, 59
    highway design and, 61
    search protocols and, 65
    for traffic, 62, **116**
    train timetables and, 118
    for tunnels, 121
    for water supplies, 129
Mathematical Reviews (journal), 33
Mathematica (software), 17
Mathematicians of the African Dias-
    pora, 34
Mathematics of Investment (Rider and
    Fischer), 15
Math Forum, 34
Math Fun Facts, 34
MathSciNet, 33
MathTrek (Peterson), 34
matrices, 65
mattresses, 73–74
Maxwell, James, 41, **74**, 89
mazes, 68
McCulloch-Pitts Theory of Formal
    Neural Networks, 79
McCulloch, Warren, 79
McCurty, Kevin, 65

Index

McKenna, P. Joseph, 13
McLean, Malcolm, 99
Measure Master, 16
measuring time, 28
mechanical clocks, 28
memory foam, 73
memory latency, 87
Message Passing Interface (MPI), 84
meter, 113
microscopic modeling, 116
Microsoft Corporation, 86
microwave ovens, 74–75
microwave technology, 74
military code
    code talking, 90
    Enigma code, 30
    Morse code, 89
Millau Bridge (France), 14
Mindstorms NXT, 93
mobiles, 70
modular arithmetic, 28
Monte Carlo simulation, 3
moon
    lunar calendars, 18
Moore, Gordon, 86
Moore's law, 86
Morpheus Laboratory, 1
Morse code, 89
Mo Ti, 41
movies, making of, 75–77
MP3 players (MPEG Audio Layer III), 77–78
Mr. Gasket Hot Rod Calc, 16
Mumford, Lewis, 27

Nano-robots, 91
nanotechnology, 87, 91
nanotubes, 87
Napoleon Bonaparte (Napoleon III), 20
National Aeronautics and Space Administration (NASA)
    space elevators, 48, 49
National Association for Stock Car Auto Racing (NASCAR), 9
National Conference on City Planning, 27
National Council of Teachers of Mathematics (NCTM)
    Illuminations Web site, 34
    roles of proof in education and, 35
National Institute of Standards and Technology, 29

National Security Agency (NSA), 32, 37
Native American mathematics
    code talking and, 90
Nave, Jean-Christophe, 62
Navier, Claude-Louis, 4, 13, 129
Navier-Stokes equations, 4, 129
navigational clocks, 28
Neonativist theories, 84
Netflix, 67
network science, 66
networks, Internet, 65, 66
Neumann, John von, 103
neural networks, 78–80
neurophysiology, 84
Newton, Sir Isaac
    biographical information, 18
Newton's laws, 5
new urbanism, 58
Nielsen, Arthur, 80, 81
Nielsen, Henrik Frystyk, 97
Nielsen Media Research, 80
Nielsen ratings, 80–82
Nine Chapters on the Mathematical Art (Chinese text), 19
Nipkow disk, 109
Nipkow, Paul, 109
Nook (Barnes and Noble), 40
NP-Complete problems, 98
NP-hard, 120
"Nuremberg eggs", 28

Obama, Barack, 126
O'Connor, John, 34
October Revolution (Russia), 18
Ohm, Georg, 47
Ohm's Law, 47
"On Computable Numbers" (Kleinrock), 64
One Laptop Per Child, 87
Optical Character Recognition (OCR), 100
optical illusions, 82–84

Page, Lawrence, 65
PageRank (Google), 65
Panama Canal, 20
parallel processing, 84–85
patterns
    recognition of, 79
Peaucellier cell, 69
perceptrons, 79, 80
personal computers, 85–87

Peterson, Ivars, 34
Peterson, W. Wesley, 46
phugoids, 4
Piccard, Auguste, 39
Piccard, Jacques, 39
Pixar Animation Studios, 7
place-value structures, 17
plain-text, 30, 31
plasma televisions, 111
Playfair Square, 31
Ponte Vecchio (Italy), 14
Pontryagin, Lev, 54
Pontryagin's Maximum Principle, 54
portable document format (PDF), 40
Postel, Jonathan, 65
Post Office Protocol (POP), 64
Potts, Renfrey, 121
power centrality, 105
Powers, Kerns H., 110
Prandtl-Glauert equation, 6
Prandtl, Ludwig, 6
prayer wheels, 131
predicting attacks
    National Security Agency (NSA), 32
pressurization, 2
Principles and Standards for School Mathematics (NCTM), 35
ProjectCalc series calculators, 16
Project Gueledon, 23
Project Gutenberg, 40
proof, 35, 36
pulleys, 87–89

quartz crystal clocks, 29
queuing theory, 116
QWERTY keyboard calculators, 17

radio, 89–91
radio-frequency identification (RFID), 100
random access memory (RAM), 86
randomness, 72
read-only memory (ROM), 15, 86
reality, measuring, 123
Reber, Grote, 90
Reed-Solomon codes, 45
Rényi, Alfréd, 66
Revere, Paul, 30
Reynolds number, 5, 129
Reynolds, Osborne, 5, 129
rhythms, 114

Riccati, Jacopo, 19
Rider, Paul, 15, 16
Rivest, Ronald, 31
Roberts, Louis, 61
Robertson, Edmund, 34
roBlocks construction system, 93
Robonaut 2, 92
robots, 91–94
roller coasters, 94–95
Rosales, Rodolfo, 62
RSA public key system, 31, 32

sacred geometry, 69
Saint-Venant equations, 19
Saint-Venant, Jean Claude, 19
Sarukkai, Sundar, 124
satellites
    coastline mapping and, 72
    GPS and, 55, 56, 57
scanning, 52
Scheinman, Victor, 93
Schweikardt, Eric, 93
science fiction
    space ships and, 106
Science Friday (radio program), 89
Science News (journal), 34
Science, Technology, Engineering, and Math (STEM), 34
sculptures
    mobiles, 70
search engines, 95–96
Search for Extraterrestrial Intelligence (SETI), 123
Sears, Roebuck and Co., 16
Sears Tower, 101
second-generation (2G) cell technology, 25
Segway, 96–97
Seibold, Benjamin, 62
servers, 97–98
Seven Bridges of Konigsberg, 14
Shamir, Adi, 31
Shannon, Claude, 30, 86, 132
Shelley, Mary, 93
    Frankenstein, 93
shipping, 98–100
Shockley, William Bradford, 86
Shor, Peter, 87
shuffle function, 78
signal processing, 76
Simple Mail Transfer Protocol (SMTP), 65

Six Degrees of Kevin Bacon, 105
skyscrapers, 100–102
SMART board, 102–103
smart cars, 103–104
Smith, George, 43
Snow, John, 25
social networks, 104–106
Society for Technical Communication, 34
Solid-state drivers (SSD), 45
Solomon, Gustave, 45
space elevators, 49
spaceships, 106–108
spam filters, 108–109
Spartan ciphers, 30
Sporer, Thomas, 77
SR-52, 15
Stanford arm, 93
Star Trek (television show), 93
Star Wars (movies), 7
Steel and Foam Energy Reduction (SAFER) barrier, 11
Ste-One, 7
Stokes, George, 4, 129
Stonehenge, 85
Stylianides, Andreas, 36
subsonic speeds, 2
Su, Francis, 34

Taschenuhr, 28
televisions, 109–111
Tesla, Nikola, 89
TeX, 34
Texas Instruments, 15, 16
thermostats, 111–112
"thinking fluid", 9
third-generation (3G) cell technology, 25
timekeeping, 28, 29, 30
time signatures, 113–114
timetables (train), 118
time zones, 118
toilets, 114–115
Tour de France, 13
tower clocks, 28
traffic, 116–117
trains, 117–120
trajectories
    fuel consumption and, 54
Transmission Control Protocol (TCP), 98
transverse flow effect, 60

traveling salesman problem (TSP), 99, 120–121
Trieste, 39
trilateration, 55, 56
Tsiolkovsky, Konstantin, 48
Tunnel of Samos, 121
tunnels, 121–123
Turing, Alan, 63
Two New Sciences (Galileo), 69

Union Canal Company, 19
Universal Automatic Computer (UNIVAC), 86
universal language, 123–124
U.S. Coast Guard, 56
U.S. Department of Commerce, 27
U.S. Environmental Protection Agency (EPA), 130
U.S. National Bureau of Standards, 29
U.S. National Collegiate Mathematics Championship, 34
U.S. Postal Service, 99, 100
U.S. Supreme Court, 27
Uzelac, Tomislav, 77

Vasco da Gama Bridge (Portugal), 14
Veltins Arena (Germany), 9
vending machines, 124–125
Verdu, Sergio, 132
vertical curves, 61
video games, 125–128
    geometry in design, 126
    programming, 126, 127
Vigenère cipher, 31
Vinge, Vernor, 87
virtual reality, 84
visual perception, 82, 83
Volta, Alessandro, 47
volts (V), 47
von Neumann, John, 86

Walt Disney Studios, 7
Warnock, John, 40
water distribution, 128–130
water footprints, 129
Water Works Bureau, 20
Watt, James, 47
Web based communication, 33
Web crawlers, 95
Weiner, Norbert
    cybernetics, 92
Westphal, Heinrich, 73

wheel, 130–131
"White City" (World's Columbian Exposition of 1893 in Chicago), 26
Whyte, Frederick, 118
Whyte system, 118
wide-area augmentation system (WAAS), 56
Wiener, Norbert, 86
Williams, Scott, 34
Willis Tower, 101
wind and wind power, 131
windmills, 131–132

wind turbines, 131, 132
Winer, David, 41
wireless communication, 132–133
Wise, Michael, 85
Wolfram MathWorld, 34
World's Columbian Exposition of 1893 (Chicago), 26
World War II
    aircraft carriers, 2
Wright, Benjamin, 20
Wright, Frank Lloyd, 27

X-rays, 44

Yackel, Erna, 36
Yeager, Charles, 6
Yost, Charles, 73

Zentralblatt für Mathematik (journal), 33
Zentralblatt MATH, 33
Zoetrope, 6
Zoltan, Kecskemeti B., 7